INSTALLING HOME THEATER

THE COMPLETE GUIDE TO BUYING, INSTALLING AND MAINTAINING HOME THEATER SYSTEMS

by Gordon McComb

This book was developed and published by:
　　Master Publishing, Inc.
　　Lincolnwood, Illinois

Edited by:
　　Pete Trotter

Printing by:
　　Arby Graphic Service, Inc.
　　Lincolnwood, Illinois

Acknowledgements:
　　All photographs not credited are either courtesy of Author, Radio Shack or Master Publishing, Inc.
　　Appreciation is expressed to Zenith Electronics Corporation, Glenview, Illinois for its assistance.

Trademarks:
　　Dolby®, Pro Logic®, Dolby Digital® are registered trademarks of Dolby Laboratories Licensing Corp. THX® is a registered trademark of Lucasfilm. DTS® is a registered trademark of DTS.

　　Throughout this book, Master Publishing, Inc., has attempted to distinguish known or suspected proprietary trademarks or service marks from descriptive terms by appropriately inserting a trademark or service mark symbol at the first mention of the trademark or service mark. Every effort has been made to supply complete and accurate information; however, Master Publishing, Inc. cannot attest to the accuracy of this information, nor assume responsibility for its use, nor for any infringement of the intellectual property rights of third parties which would result from such use. Inaccurate use of a term in this book should not be regarded as affecting the validity of any trademark or service mark.

Copyright © 1998
Master Publishing, Inc.
7101 N. Ridgeway Avenue
Lincolnwood, IL 60645
voice: (847) 763-0916
fax: (847) 763-0918
e-mail: MasterPubl@aol.com
All Rights Reserved

Visit Master Publishing at
www.masterpublishing.com

REGARDING THESE BOOK MATERIALS
Reproduction, publication, or duplication of this book, or any part thereof, in any manner, mechanically, electronically, or photographically is prohibited without the express written permission of the publisher. For permission and other rights under this copyright, write Master Publishing, Inc.

The Author, Publisher and Seller assume no liability with respect to the use of the information contained herein.

First Edition

9　　8　　7　　6　　5　　4　　3　　2　　1

Table of Contents

		Page
	Preface	iv
Chapter 1	What is Home Theater?	1-1
Chapter 2	Program Sources	2-1
Chapter 3	The TV: Silver Screen of Home Theater	3-1
Chapter 4	Sound — The Other Half of Home Theater	4-1
Chapter 5	Hooking Up the Home Theater System	5-1
Chapter 6	Satellite Dish & "Off-the-Air" Antenna Hookups	6-1
Chapter 7	Wiring Your Home Theater System	7-1
Chapter 8	Room Design and Ergonomics	8-1
Chapter 9	Home Theater Maintenance	9-1
Chapter 10	Troubleshooting Common Problems	10-1
	Glossary	G-1
	Index	I-1

Introducing Home Theater

In 1885, at the Institute for the Deaf in New York City, French inventor Louise Aimé Augstine Le Prince was experimenting with the idea of creating moving photographic images. He designed a camera for taking such pictures, and a projector for showing them. Le Prince filed for a U.S. patent to protect his invention but, for technical reasons, only a portion of it was accepted. His patent for the "Apparatus for Producing Animated Pictures" was granted in early 1888, and is the earliest-known, serious effort to create a camera and projector for moving filmed images.

Later that same year, a far more famous inventor, Thomas Edison, filed a caveat with the Patent Office for an "optical phonograph." Edison's original idea was crude and unworkable. Eventually, he turned to a lab assistant, William Dickson, to develop a working motion picture device. By 1892 Dickson created a fully-functional motion picture camera using vertically-fed, perforated film—a camera not unlike the one Le Prince had patented four years earlier, and very similar to the movie cameras used today. Later, Edison was given almost sole credit for developing the motion picture camera and projector, even though the pioneering work was done by other men. No matter who deserves to be called the "father" of motion pictures, there is no doubt that *the movies* have made a lasting impression on most everyone's life.

The potential for success wasn't lost on Edison. He set-up the world's first movie studio, and the first commercial presentation of his "Kinetoscope" films was held in April, 1894 in New York, where patrons paid twenty-five cents to see five filmed "short subjects." The blockbusters of the time ranged from rooster fights to a man getting a shave.

Fortunately, audiences worldwide have since enjoyed a broader fare, from the famed MGM musicals of the 30s and 40s, Douglas Fairbanks "swashbucklers" and a nearly endless onslaught of Frankenstein, Dracula, and Werewolf pictures; to "serious" drama like *Elmer Gantry;* spectacles such as *Ben-Hur;* screwball comedies with favorite duos like Bob Hope and Bing Crosby, to today's "blockbuster" hits.

With an art form as important as the cinema, it's no wonder the concept of *home theater* is so popular today. It is a way to bring the movie theater experience into our homes. No need to find a baby sitter, stand in line, or pay $4 for a container of popcorn; everything great about seeing a movie in a theater is no further away than your own living room.

This book will serve as a guide for you to create a movie theater environment in your own home. It discusses all aspects of this audio/visual revolution from TV sets, DVD players and satellite dish antennas down to the speakers, wires, connectors and plugs you'll need to plan, install and maintain your home theater system.

We hope you find it helpful, and we hope you enjoy the show!

G. McC.
MP

What Is Home Theater?

The concept of *Home Theater* is to take full advantage of today's advancements in audio and video technology for in-home entertainment. At a very simplistic level, home theater combines your TV and stereo system to create an "environment" for watching movies and television shows. But true home theater is much more than that. It's about setting up the best sound system for the room, and tailoring the video system to provide crystal-clear pictures from a variety of program sources to create a home-based theater that the entire family can enjoy.

In this book, you'll learn how to design, install, troubleshoot, and maintain your own home theater system. You'll discover the options available in TV screens, sound systems, and speakers, and how to connect everything together. If you've priced full home theater systems and you're worried about sticker shock, relax: this book doesn't assume you are buying a home theater system from scratch. You'll find plenty of information on updating your current television and sound system to create a functional, enjoyable home theater.

The Typical Components of a Home Theater System

Home theaters are an amalgam of video and audio components. The major components, as shown in *Figure 1-1,* are:

- A television set (or video monitor) for watching the picture—it's the "silver screen" of your living room.
- An A/V receiver for selecting the video source to view and to properly "decode" the soundtrack for maximum enjoyment.
- A set of speakers, strategically placed around the room, that provide surround sound for the audience.

THE A/V RECEIVER

An A/V stereo receiver, such as the one shown in *Figure 1-2,* connects everything together—it is the nerve center of a home theater system. While it serves the same basic function as a standard stereo receiver in a hi-fi audio system, there are some important differences you need to understand.

First, the A/V receiver decodes the surround sound soundtrack and drives the various speakers with the soundtrack portion of the movie or television show you are watching. This is perhaps the most important job of the A/V receiver, because home theater relies heavily on surround sound and multiple speakers.

Second, the A/V receiver is able to switch among multiple audio and video sources. In addition to switching between stereo audio sources including radio tuner, CD player,

••• INSTALLING HOME THEATER •••

phono, and tape, most A/V systems also can switch between at least two video sources. These video sources must provide separate audio and video signals. Therefore, applicable video sources for connection to an A/V system include VCRs, video disc players, and most satellite receivers, but not the typical cable converter box. VCRs, disc players, and satellite systems usually have separate audio and video outputs, in addition to the regular antenna output. Most cable converter boxes can only attach to the "Antenna In" connection on your TV or VCR. (For home theater, you should use your VCR to route sound and picture from a cable box or antenna through the A/V receiver. This is helpful if you are watching a surround sound program. See Chapter 5 for more information.)

Figure 1-1. The main components of the typical home theater system include a television, an A/V receiver, and a set of speakers.

The ability of the A/V receiver to select the desired video source—such as a VCR, video disc, or satellite dish—and route it to the television is a great convenience if you have a number of program sources because you don't need to manually connect wires or press confusing switches to connect the desired programming source to your TV.

 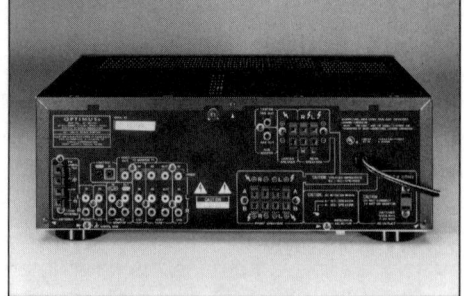

Figure 1-2. The A/V receiver functions as the video/audio selector, and provides the surround sound stereo to the speakers in your home theater system. (Optimus STAV-3690, RS #31-3040.)

1-2   What is Home Theater?

THE TELEVISION

Next comes the television. Note that "television" can cover a wide range of options, from 25-inch console models to large projection units. For the most part, the bigger the screen the better, but there are practical limitations to consider, including cost and room size. Chapter 3 provides more detail how to select the best TV for your home theater. However, as a starting point, you should consider a television with a screen no smaller than 25 inches (measured diagonally), with 27 or 32 inches preferred.

Some home video purists insist that a projection TV is required to get the full theater-at-home experience, but not everyone has the space or bank account for a quality projection unit. Your home theater will provide many hours of enjoyment, even if the TV is a generic 25-inch, direct-view model.

Perhaps as important as the physical size of the screen is the type of signal input the TV accepts. As explained in later chapters, newer televisions accept two kinds of input: RF and direct. RF inputs combine both sound and picture, modulated on a radio frequency signal (thus the "RF" nomenclature). Examples of RF signal sources include a television antenna, the "Antenna Out" connector on a VCR, and the output of most cable boxes.

Direct inputs provide only the audio or video portion of the signal, without the RF modulation signal. Direct inputs generally provide sharper pictures and cleaner sound. You will get more pleasure from your home theater system if your TV has direct inputs for audio and video. Almost all newer TVs provide direct inputs, but many older models do not. If your TV lacks direct audio and video inputs, you should seriously consider replacing it with a newer set. Though the television is perhaps the single most expensive component of the average home theater, a quality picture is a must if you are to enjoy the full benefits of your system.

THE SPEAKERS

Note the array of five speakers shown in *Figure 1-1*. Home theater shuns the single speaker that is built into most television sets and replaces it with a battery of higher-quality speakers situated around the room. The main speaker is located on or over the TV and is called the *center-channel speaker*. The bulk of the dialog, music, and special sound effects of the program you are watching comes through this speaker. Because most of the sound comes from the center-channel speaker, its quality is paramount to a good home theater system.

Flanking the TV are the *left- and right-front stereo speakers*. These augment the center channel speaker and provide a "wider" sound stage, just as they do with traditional stereo systems. By using the left and right stereo speakers, the sound seems richer and fuller. In addition to on-screen dialog, the left- and right-front speakers provide the stereo effect for music and background sounds. If your hi-fi system is located in the same room as your TV, you can probably use its speakers for the left- and right-front speakers of your home theater system.

Completing the sound stage of your home theater system are the *left- and right-surround* (also called *rear*) speakers. These speakers are positioned to either side of the seating area, or behind it. The surround speakers provide background sound for added ambiance, placing the audience in the middle of the sound stage. They are typically used to enhance sound effects, like gunshot ricochets or the whirring sound of a robot motor.

••• INSTALLING HOME THEATER •••

Though there are two surround speakers, in the typical home theater setup the sound from these speakers is not in stereo. Both speakers are fed by the same sound source and are, therefore, monophonic. This "rule" doesn't apply to home theaters equipped with digital soundtrack systems, such as Dolby Digital® (also known as AC-3), or the DTS® digital system. Digital soundtrack systems such as these call for stereophonic surround speakers. Read more about digital soundtrack systems below, and in Chapter 4.

In addition, some home theater systems add a *subwoofer* to the array of speakers. The subwoofer provides the deep bass sound used in many sound effects. More about the subwoofer in Chapters 4 and 5.

HOME THEATER STEREO

You can always tell that a program source has been encoded for special sound when you see the notation *Dolby Surround*® or *Dolby Stereo*®. This indicates that the program was recorded using the industry-standard Dolby stereo encoding technique.

To successfully decode the surround sound signal and play it through the five speakers of your home theater, your A/V receiver must be equipped with a Dolby Stereo decoder. There are two forms of Dolby Stereo decoders for the general consumer home theater market: matrix (sometimes called discrete or 3-channel logic) and Pro-Logic®. Of the two, Pro-Logic is the preferred system, and is used on the majority of A/V systems sold today. If you are in the market for an A/V system, look for one that contains Dolby Pro-Logic circuitry. It not only will faithfully reproduce a program encoded in Dolby stereo, but also will be able to synthesize the "surround sound" effect for those programs recorded in standard stereo format.

(NOTE: The traditional stereo receiver is capable of driving only a pair of speakers and lacks the additional center channel and surround speaker outputs required for home theater. Of course, you can use a regular stereo receiver and listen to the program sound through your hi-fi system, but this is not true home theater.)

> **Note:** The term Dolby also is used for other types of sound processing, the most familiar being noise reduction for audio tapes where the level of background noise is reduced compared to the program signal. Video tapes, discs, and other video sources seldom use Dolby noise reduction. If you see the "Dolby" notation on a videotape box, you can assume it means Dolby Stereo, not Dolby noise reduction.

HOME THEATER PROGRAM SOURCES

"Program sources" are where you get the pictures and sound to play through your home theater system. In decades past, the only program source was an off-the-air antenna that might allow you to tune into five or six local broadcast TV channels. If there was nothing of interest on any of the channels, it was time to dig the Monopoly® game out of the closet.

Today, as shown in *Figure 1-3,* consumers enjoy at least five major program sources for their television—and home theater—enjoyment. These program sources are discussed in more detail in Chapter 2, as well as in other chapters throughout this book. The five major program sources are:

1-4   What is Home Theater?

••• INSTALLING HOME THEATER •••

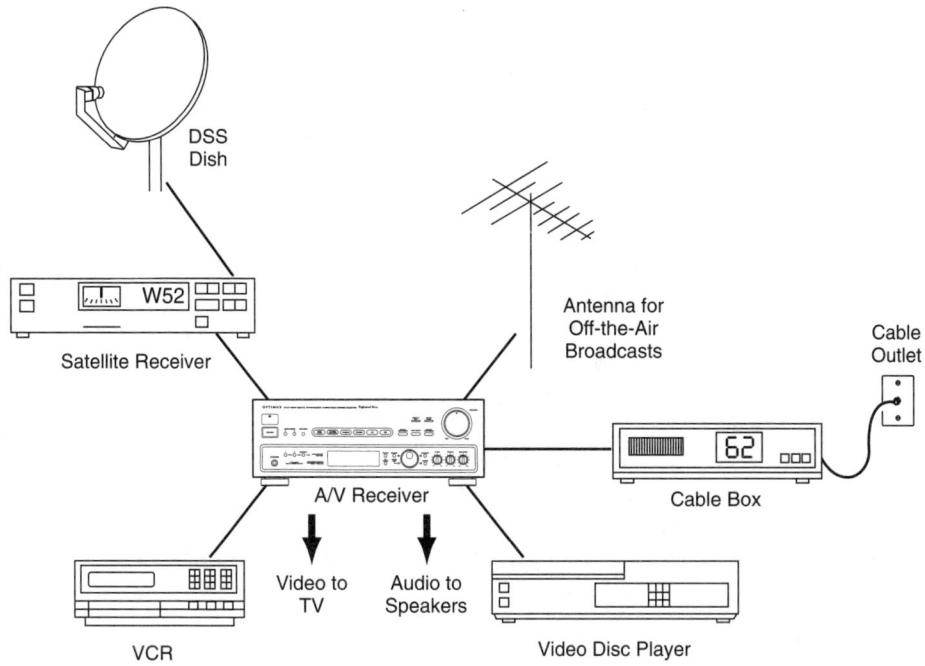

Figure 1-3. Your home theater system can choose from among five popular programming sources, including VCR, cable, and off-the-air broadcasts. The A/V receiver is typically used to select the programming source you wish to view.

- *Off-the-air broadcasts.* Local television stations continue to provide programming using VHF channels 2 through 13, and UHF channels 14 through 69. If you want to install a new outdoor antenna for your home theater system, be sure to read Chapter 6 on antenna hookups. (As a side note, the frequencies for old UHF channels 70 through 83 were reassigned by the Federal Communications Commission for cellular telephone use. If you have an older TV set that tunes past channel 69, you'll never see a legitimate broadcast on broadcast channels 70 to 83.)
- *Cable.* More than two-thirds of the United States is "wired for cable" and able to receive television programming via a closed-circuit cable system. The benefits of cable include reception that usually is better than off-the-air broadcasts, access to additional channels, and no need for an off-the-air antenna. Cable systems offer both "basic" and "premium" channels. Channels that come as part of the basic cable service usually include re-broadcasts of local TV stations, community access channels, and one or more specialty channels, such as Cable News Network® or ESPN®. Premium channels cost extra and provide optional programming, such as HBO® (movies), or Nickelodeon®.
- *Home satellite.* Once just the domain of hobbyists, home satellite television reception systems have become the hottest-selling consumer electronics product ever introduced. One benefit of home satellite television is that you have more than a hundred channels to choose from (like cable, it costs extra to receive premium satellite channels, such as HBO). Another is, you don't have to be hard-

What is Home Theater?   1-5

wired to a cable system. If you want to install a new satellite antenna for your home theater system, be sure to read Chapter 6.
- *Video cassette recorders.* VCRs provide a means to record programming from another source (off-the-air or cable, for instance) for later viewing. They also provide a means to play pre-recorded videotapes, such as movies or documentaries. The tapes can be purchased or rented, thereby giving you the ability to be your own "programming executive." This is perhaps the primary reason for the success of the VCR; it releases you from the tastes, timing and interests of some network or local broadcast station programming executive. You decide what you want to watch, and when.
- *Video disc players.* Like VCRs, video disc players let you select your own programming, which you may buy or rent. Disc players are not as popular as VCRs, mainly because they cannot yet record. Older style analog machines play 8- or 12-inch discs; the newer digital machines play 4.7-inch discs. The smaller discs tend to be less expensive and generally produce better pictures.

Other program sources exist as well, but are not as commonly used. For example, some areas offer "wireless cable," which is a form of microwave transmission of television signals. You receive the channels via a small, outdoor dish aimed at a nearby broadcasting station. And some areas use a shared or "community" antenna system, where television broadcasts are received at a master antenna, then fed via a cable to each home (this is typical, for example, in some of the older retirement communities and in some mobile home parks).

Planning For Your System

The more you plan for your home theater system, the easier it is to install and maintain it. Knowing ahead of time what is required for a good home theater setup also will save you time and money. For example, if in the planning stage you discover that your existing stereo speakers will not hold up to the demands of home theater, you may wish to chose a home theater A/V system that includes speakers. This may save you money. You may also discover that your television set, now showing signs of age and lack of features, is in need of replacement. You may consider an all-new complete home theater package, including A/V system, speakers, and TV monitor. Conversely, you may already have most of the components you need for a quality home theater system, except for the rear speakers and a subwoofer. In this case, you will keep what you have and add the speakers as needed. Whether you purchase an all-new home theater system, or upgrade your existing video system, following are some points you'll want to consider as you plan for your new system.

BUYING AN ALL-NEW SYSTEM

Perhaps the most straightforward road to home theater is to buy a complete system, including the television, A/V receiver, and all the speakers. Usually, but not always, the components of the system are built by the same manufacturer and designed to complement one another. When buying a complete system, you can be better assured that the speakers will be rated for the power output of the A/V receiver, for example. And, quite often, the manufacturer will offer a reduced price for the entire system compared to the cost of individual components purchased separately.

••• INSTALLING HOME THEATER •••

MIXING AND MATCHING

There is no "rule" that says you must purchase a complete system from one place. You may decide you like the TV made by Company A, the A/V receiver sold by Company B, and the speakers from Company C. As long as the components are rated for one another—the speakers can withstand the power from a high-wattage A/V receiver, for example—you can easily mix-and-match components. Consider, however, that components purchased separately may cost more than the components included as part of a unified system.

While you can freely mix TVs, VCRs, DVD players, and A/V receivers, when it comes to the speakers it is recommended that all of the speakers in your home theater system come from the same company. In addition to technical issues of the speakers themselves and enclosure design, the speakers will share similar jacks to attach the speaker wires or cables. Many speaker companies use distinctive designs for the front grill and cabinet enclosures and your system will appear better integrated if all the speakers share the same overall design.

Of course, this is not to say that you cannot mix speakers from different manufacturers. For example, you may prefer a certain subwoofer offered by a company other than the one that built the other speakers in your system. In that case, you can strive for a close match in design. For example, if your other speakers have wood-tone enclosures, you may opt for a subwoofer that also has a wood-tone enclosure, instead of a black-metallic enclosure. The choice is up to you, and the general decor of the room.

UPGRADING AN EXISTING TV OR SOUND SYSTEM

It's not always economically feasible to replace your existing video and sound components to create an all-new home theater system. For example, your television set, though a few years old, may be perfectly suited to being a member of your home theater team. There is little reason to replace it with a new model just to have the latest and greatest in home theater gear.

In judging the viability of your existing video and audio components, consider not just their age or exterior appearance but their adaptability to being used in home theater. Some specifics:

- The television set should have separate audio and video inputs, as shown in *Figure 1-4,* not just an "antenna in" input. These separate inputs will provide the best pictures possible from your various program sources, such as a VCR, satellite dish, or video disc player.
- The TV screen should be not less than 25 inches in size (measured diagonally). An even larger screen (27 to 32 inches or more) is preferred, as many of the quality tapes and discs for home theater are presented in "letterbox" format. This format shows the film in its original wide-screen, theatrical format. You see all the action the film makers intended, but the height of the overall image is reduced, so you need a larger-screen TV to adequately see the picture.
- Your stereo receiver should be capable of surround sound decoding, preferably Dolby Pro-Logic or Dolby Digital (the older Dolby "matrix" surround sound technology will do in a pinch). Your older-model stereo receiver lacks the features needed for home theater and should be retired, or moved to another part of the house where you may want a hi-fi system for audio listening.

••• Installing Home Theater •••

- The home theater experience depends on being *surrounded* by sound. This requires the use of at least two rear-channel speakers, and preferably the addition of a center speaker placed on or near the television set. Though not recommended, in a pinch you can use the built-in speaker of the television set as the center speaker, but the overall sound quality of your system is likely to suffer.
- Your stereo A/V receiver should provide the capability to select the video input for display on your television. However, if it does not, you can add a separate video/audio selector that will do the same job. See Chapter 5 for more information on video/audio selectors.

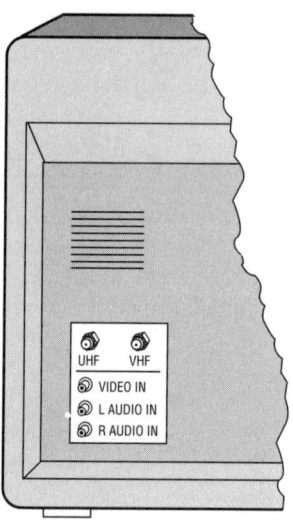

Figure 1-4. For best results, your TV should have separate audio/video input jacks, not just an "Antenna In" terminal or connector. The separate audio/video jacks provide better picture quality.

Practical Knowledge

If you've connected your VCR to your TV, or your speakers to your stereo system, then you have all the skills needed to install at least a basic home theater system. Installing a home theater system involves the same procedures. Prior knowledge of electronics, electricity, or electrical wiring is not required, though it is recommended if you wish to install a more complex system. Obviously, this book cannot address all of the fundamentals of electronics and electricity, and it assumes that the reader has a certain level of basic expertise. If you would like to know more about electricity and electronics, purchase *Basic Electronics,* by G. McWhorter and A. Evans, published by Master Publishing, Inc., available at Radio Shack (RS# 62-1394).

If you do not have the skills needed to plan and install your own home theater system, consider obtaining the services of a professional installer. Professional home theater installation is available through many specialty electronics and home appliance stores. Ask for a quote, and be sure to get references. For your own protection, the installer should be bonded and insured. And, if your state or municipality requires it, the installer should be licensed or certified.

1-8   What is Home Theater?

••• INSTALLING HOME THEATER •••

TOOLS NEEDED FOR THE JOB

With only a couple of exceptions, no special tools are required to install and use the typical home theater system. An assortment of basic tools, including screw drivers, wrenches, pliers, and so forth, will form the core of your home theater toolbox. If your home theater requires special installation (wiring through walls, for example), you will want an assortment of hand tools for working with electrical wiring, including wire cutters, insulation strippers, and wire crimpers.

While specialty tools are not an absolute requirement, a few are highly recommended. We suggest you add the following two items to your toolbox if you do not already own them (hands-on use of these tools is covered in the chapters that follow). Both of these tools are available for under $40 each.

- VOM meter *(Figure 1-5a)*. The volt-ohm (VOM) meter is used to take electrical measurements. It serves double-duty: it registers the voltage in a pair of wires or a circuit, and it measures electrical resistance in a wire or circuit. VOM meters have other applications as well. A common use is checking the "continuity" of a circuit. For example, you can check to determine if the conductor of a wire (or cable) is broken.
- Sound level meter *(Figure 1-5b)*. A sound level meter allows you to take measurements of the actual loudness of sound in a room. The meter has numerous uses, including helping you find "dead spots" in the acoustics of a room, and determining if the sound level is equal for all speakers in the system.

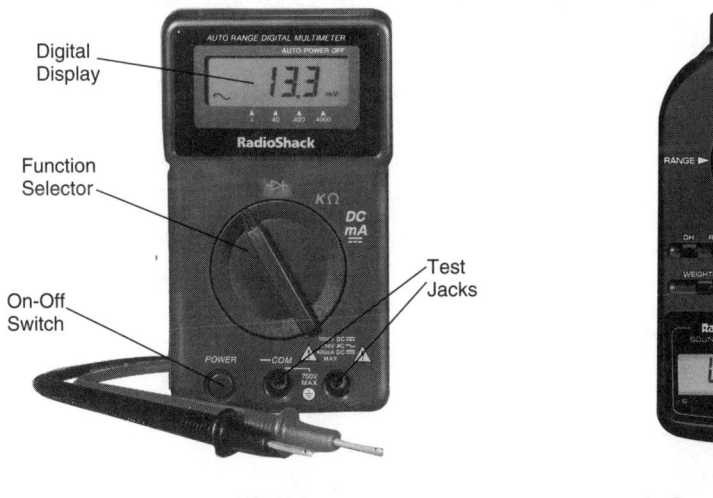

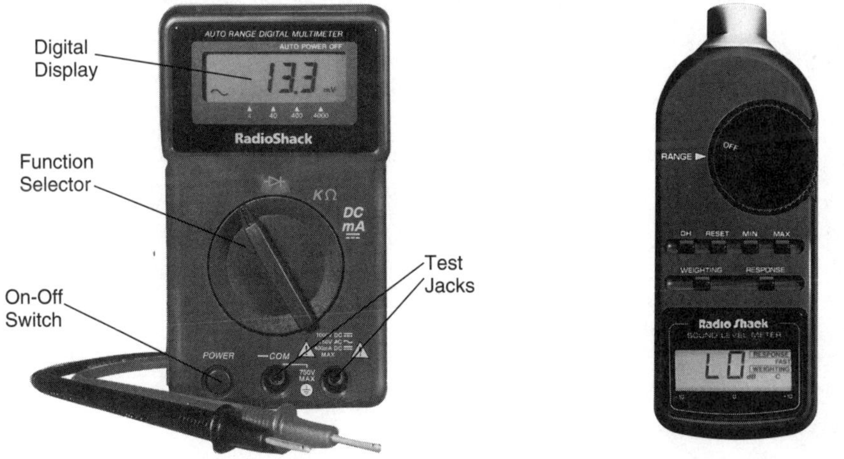

a. VOM Meter   b. Sound Level Meter

Figure 1-5. a.) The typical VOM meter is used to measure continuity as well as voltage. Basic meters start at about $10; the better models incorporate more advanced features. b.) The sound lever meter allows you to measure the relative loudness of sound at any point in the room. It is the ideal tool for setting up and troubleshooting your home theater system. The model depicted uses a digital display; less expensive (but still highly accurate) sound level meters use an analog mechanical "needle."

What is Home Theater?   1-9

Safety Always Comes First

Under normal conditions, commercially-made home theater systems are safe to install and use. The latest components follow strict safety guidelines of both the manufacturer and Underwriter's Laboratory (UL). However, care must still be exercised when installing and using a home theater system, as they are powered by 117 volt ac outlets. A malfunctioning component or faulty wiring could expose you to a shock hazard, so always use caution. This is especially important if the sound system is used outdoors, regardless of weather conditions.

In addition, many home theater systems are augmented by an outdoor antenna, either the standard VHF/UHF type, or a satellite antenna. Exercise care when installing an outdoor antenna; there is a danger of falling off the roof or a ladder. There also is potential danger of electrocution, should the antenna touch overhead electrical wires, or be struck by lightning. See Chapter 6 for additional precautions and caveats when installing an outdoor or satellite antenna.

Unless you have specific training and experience, you should *never* open the case or cabinet of an ac-powered home theater system component. Doing so may expose you and others to potentially lethal voltages. These components contain internal power supply circuits that convert 117 volt ac to high-voltage direct current.

Though *extremely rare,* there is a danger that a faulty or improperly wired sound system could cause serious injury or death to anyone touching one of the components. Be sure to read the sections at the end of Chapter 5 for some simple tests you can perform to ensure that the ac wiring to your home theater system is operating properly and is safe to use. You will need a VOM meter to conduct these tests.

Moving On

In this chapter you learned what a home theater system is and how it is used. You also learned the practical knowledge and skills you need to successfully install, operate, and maintain a sound system. In the next chapter you'll learn about the program sources you are likely to use in your home theater system. It also explains the different forms of video signals provided by the various components and how to achieve the best picture quality for your viewing pleasure.

Program Sources

A home theater system is a collection of electronic components to display pictures and produce sound with theater-like ambiance. No home theater is complete without at least one program source, and this chapter discusses the common sources used in a home theater system. They include:
- off-the-air antenna for local TV broadcasts
- cable TV systems
- satellite dish antennas and receivers
- VCRs, and
- video disc players

This chapter also discusses some important technical details about how these program sources interconnect with the other components of your home theater system, including the types of cables and connectors that are commonly used. And to help you achieve maximum performance from your theater setup, vital "background" information is included that explains the difference between RF and direct audio/video signals, video resolution, and the three means by which video signals are routed from component to component in a typical home theater system.

Antennas for Off-the-Air Broadcasts

Television antennas have been a part of the skyline since the 1940s. Look out over most any residential area and you'll see houses sprouting aluminum TV antennas of various sizes and shapes. The proliferation of cable TV has partly done away with outdoor television antennas, but many houses still use a TV antenna of some sort. Sometimes, the picture quality of local television broadcasts is higher with an off-the-air antenna than through a cable system.

Receiving off-the-air transmissions is easy: just set up your antenna to catch the radio waves and dial in the right channel. If the television station's transmitting tower is nearby, you can probably connect a small indoor antenna to your TV, rather than going to the time and expense of installing an outdoor model. Should the transmitting tower be located some distance from your house, however, you will probably need to use a larger antenna installed on the roof of your house or in the attic. Chapter 6 discusses outdoor antennas and satellite dish systems in detail.

Cable Systems

Cable television is a form of closed-circuit TV. Broadcasts from local television stations, as well as those from satellite transmissions, are received at a central site and sent over a cable through your neighborhood and into your house. Depending on the cable system, you may have access to anywhere from a handful to over 100 channels.

Most cable systems require the use of converter "boxes," which are devices specially designed to tune into the channels for that cable system. You tune your TV or

VCR to channel 3 or 4 and then use the cable box to dial the channel you want to see. The cable box receives the channel, processes the signal as required, and sends it on channel 3 or 4 for viewing on your TV.

Most cable systems *scramble* some or all of their channels. You must use the cable box on those systems where all channels are scrambled, or even the "basic" channels will be unviewable. Other cable systems scramble only the premium channels. The remaining channels are "in-the-clear" and you can use any cable-ready TV or VCR to tune directly to these channels that you wish to see, bypassing the cable box (see *Figure 2-1*).

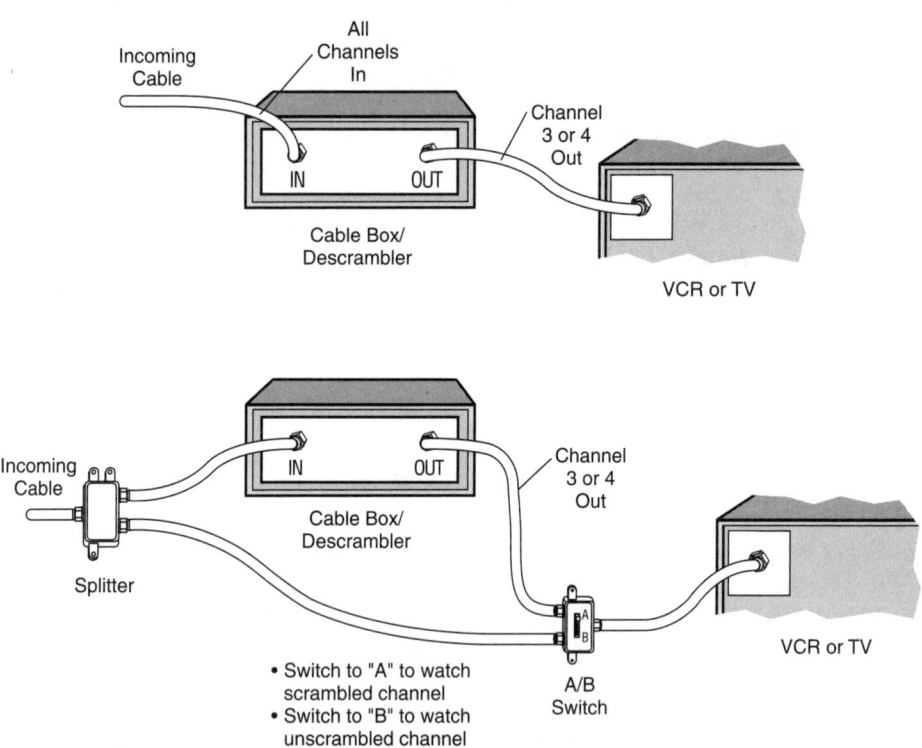

Figure 2-1. Use a cable box to tune scrambled or unscrambled channels on your cable system. For cable systems that do not scramble basic channels, you can use a cable-ready tuner in a TV or VCR to view these channels.

Video Cassette Recorders

The Sony® Betamax™ was the first widespread, commercial video cassette recorder (VCR) to hit the market. It was initially released in 1975, with production models available in 1976. But the Beta VCR was far from the first home video recorder.

Early home video pioneers developed crude systems as far back as the early 1960s. For example, in late 1964, you could purchase a video recorder kit that stored about 30 minutes of black and white programming on a single 11-inch reel of quarter inch tape (the tape speed was an astonishing 150 inches per second, some 40 times

2-2   Program Sources

faster than audio recording). By 1969 and 1970, several companies—including RCA, Sony, Magnavox, and others—were demonstrating various technologies they hoped would spark the beginning of a home video revolution. In fact, it was during this time that Sony publicly demonstrated its video cassette recorder—the "Beta" and Betamax names would come much later.

Today, VCRs, like the model shown in *Figure 2-2,* have become a common appliance in most homes. More than 80% of all homes in the U.S. with a television also have a VCR. The family VCR is used to watch rented or purchased pre-recorded videos and to tape programs off the air. The video cassette recorder has become synonymous with television watching, allowing anyone to become his or her own programming executive. Don't like what's on any of the channels? Just drop in your favorite old movie, kick back, and enjoy.

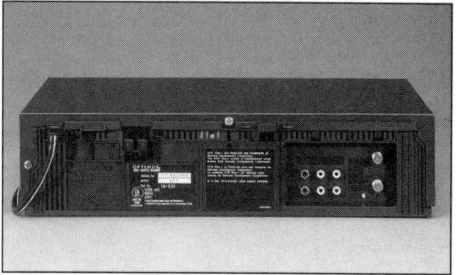

Figure 2-2. VCRs are found in most homes in the U.S. The VCR lets you watch pre-recorded videos or record shows off the air. (Optimus 63, RS #16-633.)

In the 20-plus years since their popular introduction, VCRs have gone through many changes. For a time, you had to decide whether to buy a Beta- or VHS-format VCR. Today, Beta VCRs are used only for high-end applications and most pre-recorded video is available only in the VHS format.

VHS decks are available in two general versions: standard and Super VHS. Both use the same size cassettes, but one provides higher picture quality than the other. Standard VHS plays all the pre-recorded videos. Super VHS plays the pre-recorded videos, but also records higher quality images on specially-formulated video tape. Super VHS is most useful when taping programs off-the-air and when duplicating videos taped with a camcorder.

FEATURES YOU'LL FIND

Common features of modern VHS recorders include built-in programmable timers for recording shows while you're away, automatic tracking to provide the clearest possible picture, an infrared remote control, recording at 2-, 4- and 6-hour speeds, and cable-ready tuning so that you can tune into *unscrambled* cable channels without a separate cable box. Additional features found on some models include:

- Additional video heads for improved picture quality at all recording speeds (look for a "4-head" model).
- Hi-Fi audio for better sound.
- Clear-screen special effects, including slow motion and freeze frame.

- On-screen menu display for easier setup.
- Built-in VCR Plus+ for easier programming—just enter the "code" for the program you wish to record and the rest is done for you. The program codes are published in *TV Guide* and many other television guides.
- Commercial Advance for skipping past one or more commercials.
- Automatic setting clock (the time is obtained by the VCR from a local broadcasting station that provides the local time as part of the television signal).

SATELLITE DISHES

For decades television networks have used satellites—hovering in geosynchronous orbits some 23,000 miles above the earth's equator—to relay programming from one part of the country and world to another. In the 1970s, HBO® (Home Box Office) began using satellites to instantaneously relay its programming and made theatrical movies available to the thousands of cable companies. Initially, HBO used traditional satellite communications that could be received by anyone with a satellite dish and associated receiver. This created a "home brew" satellite TV business where homeowners received network television, HBO movies, and any other programming beamed via satellite free of charge.

Within a few years, however, satellite signals were scrambled to prevent free home reception of satellite programming. Rather than killing this emerging form of home entertainment, however, a new form of programming was created for anyone who wanted more viewing options than those provided by an off-the-air antenna or cable system. Program providers like HBO soon found business in selling their channels to owners of satellite dishes.

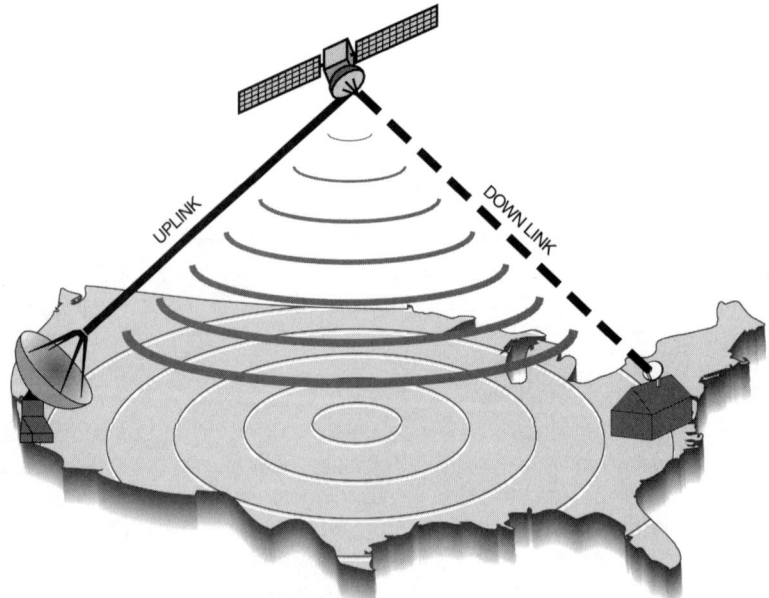

Figure 2-3. Satellite transmission of TV signals make it possible for everyone to receive hundreds of channels of programming—whether they live in a major city or on the remotest ranch.

Source: Installing TV Video Systems ©1996 Master Publishing, Inc., Lincolnwood, IL

••• INSTALLING HOME THEATER •••

TYPES OF SATELLITE SYSTEMS

Today there are two general forms of home satellite reception systems:
- The older "big dish," or Home Satellite System (HSS), system—using 7 1/2-foot or larger dishes, receives programming from traditional satellites—similar to the ones used by cable companies and television networks. The dish antenna is used along with a receiver/descrambler—called an IRD (for Integrated Receiver/Descrambler)—to watch either scrambled or unscrambled programming. Given a compatible IRD, HSS can receive three types of programming: unscrambled, analog scrambled, and digital (scrambled or unscrambled) signals. You must pay a monthly or yearly fee to receive scrambled channels. Two different communications bands are used for HSS: most programming is available on the 4 gigahertz (GHz) C-band; some additional programming is available on the 12 GHz Ku-band.
- The newer, Direct Broadcast Satellite (DBS) systems use smaller dish antennas—usually 18 inches—and a companion IRD (see *Figure 2-4*). All programming is supplied by a single satellite specifically designed to provide video programming to home users. The digital signals are encrypted and compressed, so you must pay a monthly or annual fee to receive them.

Digital satellite systems are often referred to by the name of the company that sponsors the satellite that delivers the programming to individual subscribers. A popular brand is the Digital Satellite System, or DSS (this system uses programming provided by two companies, DirectTV and USSB). Other, incompatible, digital satellite systems go by the names of EchoStar™ (also known as DISH Network), and PrimeStar®. Additional programming provided by competing digital satellite systems is expected to develop in the years to come.

You can read more about satellite systems, how they work, and how to install them in Chapter 6.

Figure 2-4. Digital satellite systems—also called DBS for Direct Broadcast Satellite—are easy to install and use. There are many systems to choose from, each offering its own proprietary programming.

Program Sources   2-5

••• INSTALLING HOME THEATER •••

VIDEO DISC PLAYERS

The video disc player, like the DVD player shown in *Figure 2-5*, is considered the finest program source for a home theater system. Video disc players provide the best quality video and sound reproduction. On a comparative basis, video discs offer up to 480 lines of horizontal resolution; compared to 230 to 250 lines provided by most VCRs, and 275 to 300 lines of resolution provided by a digital satellite system. That's an increase in sharpness of more than 50%. The concept of video resolution is explained shortly in the section entitled "Understanding Video Resolution."

On the sound side, laser video discs can be encoded with digital audio, providing sound quality as good as — and even exceeding — the best audio CDs. The latest movies on video discs are encoded with Dolby Digital or DTS digital stereo sound, the same sound system formats used in many better movie theaters.

Finally, because of the design of the laser discs themselves — using laser light to read minute patterns of information encoded in the disc — they will last indefinitely, assuming they are handled and stored properly.

Figure 2-5. Video disc players provide high-quality picture and sound reproduction using laser light to read discs encoded with video and sound track. The newer DVD players, like this one, play smaller 4.7-inch digital discs. Older models read analog laser discs. "Combi" models are available that read both the small digital and large analog discs.

TWO TYPES

Laser disc players are available in two distinct models: analog and digital. Analog laser disc players use the older, double-sided 8- or 12-inch analog laser discs — the 12-inch discs are more common. While this technology dates back to the late 1970s, analog laser discs are still popular today. The disc stores the video information in analog format; the audio portion can be stored in either analog or digital format, or both.

Digital video discs, or DVD, play smaller 4.7-inch compact discs. The discs can be single- or double-sided, though most current DVD discs are single-sided. Both the video and audio information is stored in digital format. As with analog laser disc players, most DVD players can also play music CDs.

Neither analog laser disc nor current DVD provide a means to record your own programming on disc (however, recordable DVD is in the works). This means that your laser disc player is designed as a playback-only machine. You should determine if a laser disc player is suitable for your needs based on the movies and other programs available on the laser disc format.

2-6   Program Sources

••• INSTALLING HOME THEATER •••

Which is better—analog or digital video disc? At first blush, it might appear that the larger-format, analog laser disc, which stores about an hour of video on a single side of a 12-inch disc, would offer better pictures than a DVD disc. By comparison, DVD stores over two hours of video and audio on a single side of a 4.7-inch disc. Generally, however, the DVD picture quality is better, despite the fact that the image is digitized and compressed. The MPEGII digital technology used with DVD helps preserve picture quality, even though the signal is compressed. All things considered, most consumers favor the DVD format, in part because the discs are smaller and easier to manage.

DVD REGION CODES

An added feature of DVD is that each disc is coded with a particular region. Regions are used to ensure compatibility of discs with players sold in a particular country (television standards are not always the same from country to country). For this reason, you cannot usually purchase a disc in, say Japan, and use it on a DVD player purchased for use in the United States. Therefore, when buying DVD discs, make sure they can be played on the DVD machine you own.

THE DIVX "SUBFORMAT"

A variation of the DVD format is DIVX, a special kind of low-cost DVD disc that you rent for a limited time rather than buy. DIVX discs cannot be played on ordinary DVD players, because the information stored on them is encrypted. When used with a DIVX player, a DIVX disc will allow two days of unlimited viewing of the disc; if you want to watch the disc some more you must purchase additional time. It should be noted that not all movie studios support the DIVX format. As of this writing, only a small number of movie titles are planned for release on DIVX.

COMBINATION PLAYERS

A number of analog laser disc players are "combi" units capable of playing digital and analog laser discs, as well as ordinary audio CDs. So, if you already have a collection of analog laser discs, you can still jump on the DVD bandwagon with a combination LD/DVD player. These combination players accept the following disc formats:
- 8- and 12-inch analog laser discs
- 4.7-inch DVD discs
- Standard 4.7-inch audio CDs

When you insert a disc, the player automatically detects which type you are using and switches to the appropriate playback mode.

SIGNAL TYPES AND INTERCONNECTIONS FOR YOUR HOME THEATER SYSTEM

Thus far in this chapter, we have discussed the primary program sources available for home theater. The remainder of this Chapter provides information that will help you understand how to obtain maximum performance from your home theater system. The concept of *video resolution* and its importance in deciding which program source to use in your home theater is discussed. The difference between RF and direct audio/video signals is explained. The methods of connection and cable types to obtain the best possible video resolution also is discussed.

••• INSTALLING HOME THEATER •••

UNDERSTANDING VIDEO RESOLUTION

Today, makers of video gear are putting greater emphasis on delivering the best possible picture. But saying that a picture is "clear" or "very clear" doesn't mean much, because what is a sharp, distinct picture to one person may appear washed-out and fuzzy to another. *Video resolution* is an important topic because it is now used as a major selling point for a number of home theater components, especially large-screen TVs, DVD players, and VCRs.

Fortunately, picture clarity—known as resolution—can be objectively measured using the proper equipment. The resolution of the picture simply indicates the smallest possible detail that your TV, VCR, or other piece of equipment can reproduce. The resolution of a television signal depends on two basic factors: the *scan rate* of the video circuits, and the *bandwidth* of the incoming signal. Please note that the following discussion is based on analog, standard-definition television. Future televisions based on compressed digital signals, now being developed, will provide improved, higher-definition picture resolution.

SCAN RATE

First, let's take a look at scan rate. You probably already know that a TV picture is made up of many horizontal lines stacked one upon another. These lines don't appear instantaneously on the screen, but are "drawn" by an electron gun inside the picture tube. The beam from the electron gun starts at the upper left corner of the tube (as you look at it from the outside), and progresses from left to right and from top to bottom. When it reaches the bottom right corner of the tube, the beam is momentarily turned off and the gun is re-aimed at the upper left corner. Then the procedure begins again.

The scan rate is directly tied to the number of lines drawn on the tube. In one-thirtieth of second, a TV in North America will draw 525 lines across the picture tube (actually, fewer than 525 lines are visible because some of the lines are used for timing purposes or carry special information like closed-captioning). Put another way, in one full second, the TV will draw 525 lines on the tube 30 times, for a total of 15,750 lines. Thus, the specified scan rate for a TV used in North America is approximately 15,750 Hertz (cycles per second).

BANDWIDTH

Bandwidth is the frequency range of the incoming television picture. You know that music sounds better when the frequencies heard encompass the full 20 to 20,000 Hertz range of human hearing. Likewise, television pictures are sharper when the frequency range is increased. You can better understand why if you look again at the scan rate of the TV. The electron beam scans the face of a tube 15,750 times each second. As it moves, the intensity of the beam must change in order to make an image on the screen. As the intensity increases, the picture gets brighter. As the intensity decreases, the picture gets darker.

If the video signal transmitted to the TV has a low bandwidth—say, two megahertz (two million cycles per second)—the beam intensity will change rather sluggishly. That means you will see fewer distinct edges and shapes on the screen. But increase the bandwidth to three or four megahertz, and the beam can alternate between high and low intensity much faster. This provides sharper transitions between light and dark segments of the screen, and that means higher resolution. Given the proper control circuits, the electron beam in a TV set can turn off and on in just a few millionths of a second, so it's unlikely that the electron gun will ever lag behind the incoming video signal.

VERTICAL VS. HORIZONTAL RESOLUTION

Video is comprised of two kinds of resolution, determined by scan rate and bandwidth. The scan rate determines what is known as *vertical resolution;* bandwidth determines *horizontal resolution* (see *Figure 2-6*). Although both are measurements of the resolving power of a television system, they are different. Vertical resolution is locked to the scan rate of the TV set, so there's no way to improve it other than to redesign the television circuits and broadcasting standards. In fact, most manufacturers don't even list vertical resolution in their product specifications sheet since all consumer gear is rated the same.

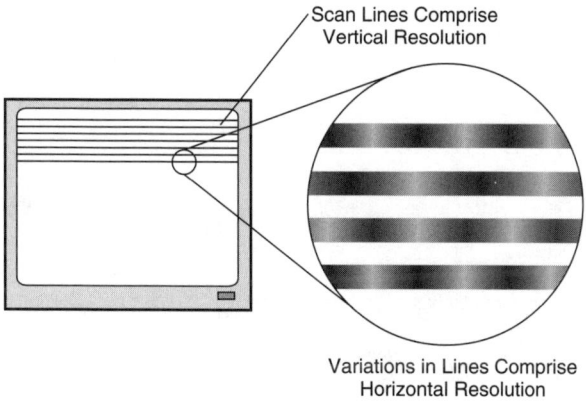

Figure 2-6. Vertical resolution is the number of scan lines traced on the front of a TV screen. Horizontal resolution is the ability to display fine detail along the horizontal axis of the screen. Vertical resolution is set by the specifications of our broadcasting standards; horizontal resolution differs depending on the program source.

The term "horizontal resolution" often causes confusion among buyers and sellers of video equipment because of the way it is physically measured. The lines on the resolution chart are drawn vertically, so some manufacturers specify horizontal resolution as "XXX vertical lines-per-inch," where XXX is some number.

Horizontal resolution is a measurement of the reaction time of the TV circuits and electron gun (or in a VCR, the tape and video heads), and determines the overall clarity of the picture. Horizontal resolution is most appropriately stated in lines-per-inch. To measure horizontal resolution, manufacturers use a standardized resolution chart that is made up of a series of patterns, or "wedges." These patterns are little more than lines spaced a certain distance apart.

Like an eye-chart at a doctor's office that tests the smallest letter you can read, the resolution chart determines the smallest pattern discernible on the TV screen. For example, conventional VHS VCRs can resolve no more than about 250 lines-per-inch, whereas a Super VHS deck can resolve 350 lines. A DVD player can resolve up to 500 lines-per-inch. Of course to see the difference, your TV must also be able to achieve the increased resolution.

What Makes Up RF and Direct Video Signals

RF (radio frequency) signals are the type received by an off-the-air television antenna, signals on cable TV, and signals received on a satellite dish. RF signals are composed of both the visual and audio parts of the program, along with a "modulating carrier"—an electrical wave that increases the frequency of the signal for broadcasting purposes (see *Figure 2-7*). RF signals can have different frequencies, hence the variety of individual channels that can be received by a television set at any one time. RF signals also are used in home video systems since they operate under an established standard and all TVs can receive them. Typically, the channel 3 or 4 output of your VCR is an RF signal that is sent to your TV set through the antenna terminals.

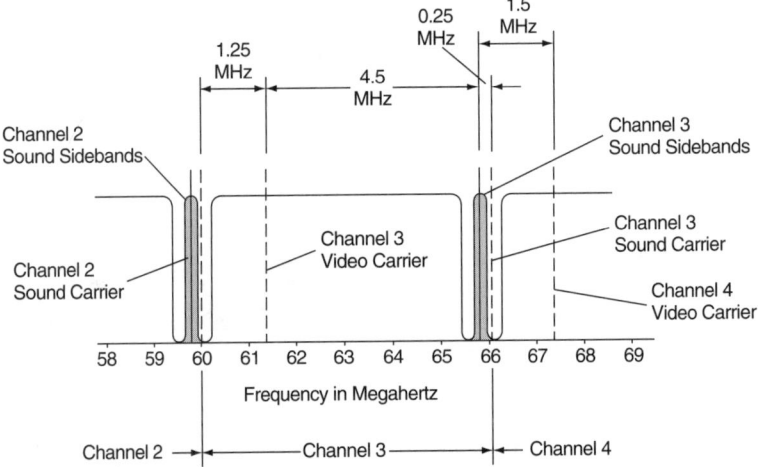

Figure 2-7. Television RF signals are composed of separate video and audio signals that are combined and placed on a higher radio-frequency carrier (modulated) for broadcasting. The frequency of the carrier determines the channel. Your TV set selects the channel for viewing and demodulates the signal, removing the carrier, and separating the baseband video and audio signals. As shown here, Channel 3 uses a total of 6 MHz of bandwidth—the video portion of the signal occupies 5.75 MHz and the audio signal occupies 0.25 MHz of bandwidth. The modulation/demodulation process reduces picture quality so, whenever you can, you should use the separate video and audio inputs in your home theater system.

Source: Antennas ©1996 Master Publishing, Inc., Lincolnwood, IL

A direct video signal is just what its name implies: video only. Video-only signals, also called *baseband* or *composite* video, lack any type of audio or carrier signals. Composite video is only used in closed-circuit TV systems since it cannot be broadcast, nor is it capable of discrete frequencies or channels.

Why is the difference between video and RF signals so important? Picture quality. When a video signal is modulated up in frequency to be carried by an RF signal and then demodulated to be shown on your TV, the quality of the video suffers. So, whenever possible, your home theater system should only use direct video signals to obtain

2-10   Program Sources

the best picture quality. RF signal sources cannot always be eliminated, however, so you cannot avoid RF completely. For example, off-the-air and cable reception is via RF.

The Forms of Direct Video Signals

There is no single type of direct video signal used with home theater. As you plan and install your home theater system you may be confronted with a dizzying array of choices for connecting program sources to an A/V receiver and TV. This section describes the four forms of video signals you are likely to encounter in your home theater system and how they differ.

Note that audio signals are fairly standard in home theater. Therefore, only one type of audio signal input or output is offered on home theater components. Audio signals that are not modulated onto an RF carrier are known as *baseband audio*.

COMPOSITE VIDEO

A color TV broadcast actually is three telecasts in one, displaying varying amounts of red, green, and blue in order to render a color image on the screen. The eye "melds" these three primary colors together to form the millions of colors in the TV picture on your screen.

Broadcast television is not composed of separate red, green, and blue signals, however. That would require too much bandwidth to accommodate each signal channel. Rather, the image is broken down into two segments: brightness and color (synchronization signals also are included, but are not critical to this discussion).

- The brightness (luminance) signal is the black and white portion of the image that determines the brightness of any particular spot on the picture tube.
- The color (chrominance) signal is the color portion of the image that determines the hue (red, yellow, purple, etc.) and saturation (pure color rendition) of the picture.

Composite video provides high-quality pictures and is the most common form of video input/output provided on VCRs, TVs, and other home theater gear. While composite video yields good results, it is not perfect. The reason: eventually the color and luminance signals must be filtered and separated as they are routed to appropriate circuitry inside the TV or VCR. This combination — and eventual separation — of the luminance and chrominance signals can produce artifacts, which impair the quality of the image. Such artifacts include "dot crawl" where fine, colored detail seems to shimmer on the screen. It is noticeable, for example, when the TV shows someone wearing a herringbone or fine checker-board suit.

S-VIDEO

S-Video, provided on higher-end home theater gear, offers an alternative video connection. S-Video (sometimes called Y/C Video) keeps the luminance and chrominance signals separate; therefore they don't have to be divided inside the TV or VCR. S-Video improves resolution, provides better color accuracy, and helps avoid "dot crawl" and other artifacts.

S-Video is sometimes confused with Super VHS; the latter is a sub-variety of the VHS tape system that provides higher overall picture quality. Super VHS and S-Video are not the same, though S-Video connections are always found on Super VHS video

••• INSTALLING HOME THEATER •••

cassette recorders. You will find S-Video connections on a wide variety of better quality video components, including VCRs, TVs, A/V receivers, DSS satellite receivers, and video disc players.

COMPONENT VIDEO

Component video is a relatively new development in home video, designed to improve on the features of S-Video. Component video separates the video signals three ways, instead of just two (luminance and chrominance) as in S-video. The three component video signals are:
- Y/R—Luminance and red channel
- Y/B—Luminance and blue channel
- Y—Luminance channel only

These three signals are then processed electronically to arrive at the separate luminance and chrominance signals used internally within a TV or VCR. Component video is most commonly found on DVD players and other high-end home theater equipment. Unless your TV is fairly new, it is unlikely that it has a component video connection.

RGB VIDEO

Though not common in consumer home theater systems, a fourth type of video signal is RGB. RGB video is comprised of three completely separate signals, each carrying information for the primary red, green, and blue components of the picture. In most RGB systems, one or two additional wires may be used for luminance only and/or synchronization signals.

RGB video provides the best overall picture quality because the individual signals are kept separate. The downside to RGB video is that components that support it tend to be very expensive. You are likely to find it only with very high end video gear, including high-quality, large-screen projection TVs.

TYPES OF CONNECTORS AND THEIR USE

There are three major types of connectors used in the typical home theater system: F-connector, phono, and four-pin DIN, as shown in *Figure 2-8*. With each, the male portion of the connector is referred to as the "plug" and the female portion is referred to as the "jack." Together, the plug and jack form the complete connector.

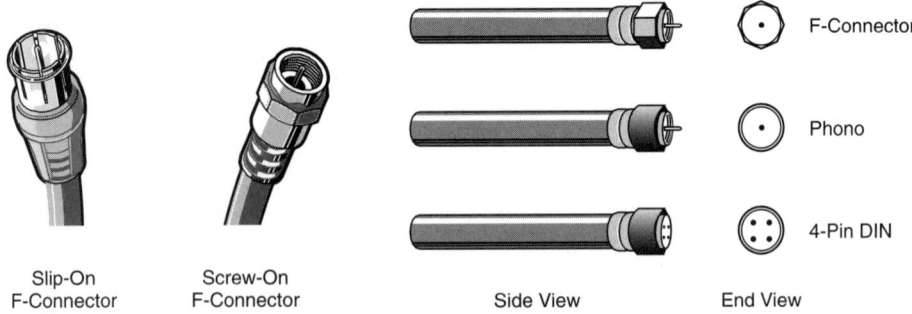

Slip-On F-Connector   Screw-On F-Connector   Side View   End View

Figure 2-8. F-connectors are used for RF signals. Phono connectors are used for video- and audio-only signals, as well as other applications in home theater. The four-pin DIN plug is used for S-Video connections.

2-12   Program Sources

F-CONNECTOR

F-connectors are used on the ends of coaxial cables that carry RF signals. The connector can be the screw-on or slip-on type. The screw-on type is preferred for permanent installation of equipment, as the connector is not likely to work itself loose. The slip-on type is convenient, and ideal when you need to quickly attach and detach components to your system.

PHONO CONNECTOR

The phono plug is a common staple of home theater systems. Phono jacks are typically used for connecting baseband audio and composite video from component to component—for example, the video/audio OUT of a VCR, to the video/audio IN of a TV set or A/V receiver. Phono connectors are comprised of two electrically insulated parts: the tip typically carries the positive side of the circuit, and the sleeve is used for the ground.

Phono connectors also are commonly used for the three separate component video inputs. Rather than one phono jack for composite video, component video provides three separate phono jacks. The jacks are color coded and marked so that you are sure to connect the plugs into their proper jacks. Bear in mind, however, that phono connectors are not considered a standard for component video. Your home theater components may use a different connector type, and therefore you may need to purchase adapters or special cables to connect the system.

Phono plugs and jacks are typically color coded for easy identification:
- Yellow is used for video inputs and outputs
- Red is used for right-channel audio inputs/outputs ("R" for Right/Red)
- White is used for left-channel audio inputs/outputs

As a point of interest, the term "phono" comes from the original application for this kind of connector in early phonographs and radios. RCA was the first manufacturer to make use of this kind of connector, which is why it is sometimes referred to as "RCA-type" or "RCA phono."

S-VIDEO DIN

S-Video uses a standardized connector consisting of four terminals, surrounded by a metal barrel that also serves as the signal shield. Two of the terminals are ground connections; the other two are for the luminance and chrominance signals used with S-Video. The S-Video connector is sometimes referred to as a DIN plug, because it uses an industry-standard DIN connector. Note, however, that DIN connectors vary in their size, number of terminals, and orientation of the terminals in the connector. This prevents you from "mixing and matching" connectors. You must be sure to get DIN connectors and cables made for S-Video.

SPECIAL CONNECTORS

Some home theater systems use additional types of connectors. Among the types you may encounter are:
- Speaker terminals, found on A/V receivers and the back of many TV sets. The terminals connect the speakers to your home theater components. Most speaker terminals are either the push-in type or pop-open type. With the push-in variety,

••• INSTALLING HOME THEATER •••

you depress a lever, then insert the speaker wire into the terminal. With the pop-open type, you move the lever to the "opened" position, insert the speaker wire, then move the lever to the "closed" position. A third type, binding posts, are used on some higher-end equipment. This terminal uses threaded metal posts. You can wrap regular speaker wire around the post, and tighten the plastic screw. Some multi-way binding posts also let you plug in a speaker wire terminated with a "banana" plug.

Quick-Connect: Press or flip to open. Depress or flip open quick connect lever; insert wire.

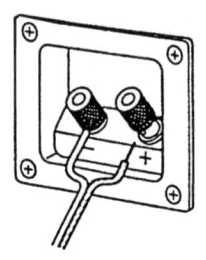

Unscrew to insert wire.

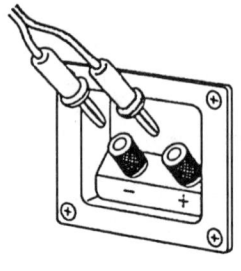

Plug-in speaker wire is terminated with banana plugs.

Figure 2-9. With a push-in speaker terminal, depress the button and insert the speaker wire. With a latch-type speaker terminal, snap the latch open and insert the wire. Close the latch by snapping it back to its original position.

- Fiber optic connections carry the video signal as an infrared light beam via a glass or plastic fiber optic cable. Fiber optic connections are found on some high-end video disc players and A/V receivers. The design of fiber optic connections varies between manufacturers of home theater components, though many manufacturers have settled on the Toslink interconnect. Many of the non-standard fiber optic connectors use a threaded plug and jack of some type.
- Dolby Digital uses either a radio frequency (RF) signal to carry its six audio channels, or a fiber optic cable (see above). Most home theater components that offer Dolby Digital audio use RCA phono plugs and 75-ohm television coaxial cable for interconnections. Since most 75-ohm cable uses F-connectors, the RF cable with RCA phono plug for a Dolby Digital connection usually is a special order item. (See Chapter 4 for more information on Dolby Digital.)
- BNC connectors look a little like phono connectors, except that have a "turn and lock" barrel that keeps the plug firmly attached to the jack. BNC connectors are used primarily in home theater systems for RGB video cables. Since relatively few consumer home theater systems use RGB video, BNC connectors are not common.

LEARN ABOUT TVS NEXT

This chapter discussed the program sources you are likely to use in your home theater system, the different forms of video signals provided by various program source components, and how picture quality is determined by the form of video signal used. It also detailed the various types of connectors used in typical home theater systems for each type of signal. In the next chapter you'll learn about television sets and the important role they play as the "silver screen" of your home theater.

THE TV: SILVER SCREEN OF HOME THEATER

Rows of television sets dot the aisles at your local electronic superstore. The same image of Michael J. Fox going "Back to the Future" in his DeLorean time machine car is flashed onto each screen. A new movie starts and the scene changes — to Arnold Schwarzenegger playing cold shoulder to Batman as the evil Mister Freeze. The overall effect is like the multi-prismatic vision of a fly — 100 Michael J. Foxes and Arnold Schwarzenneggers on 100 color TV screens.

Beyond the size of the picture tube, logic tells you that the image on each TV should be about the same, since all the sets are fed the same picture from a single digital video disc (DVD) player. Not true. For some reason unknown to you, the picture on some TVs seems sharper, brighter, and less distorted. In fact, side-by-side comparisons show a marked difference in picture quality. Why? For good reason.

The manufacturers of the latest VCRs, Super VHS decks, DVD players, digital satellite receivers, and other state-of-the-art video hardware have improved the video quality of their products, resulting in the need to create better television sets to reproduce pictures of equal quality. Television set manufacturers have been meeting this need by quietly improving their wares. But not all TV sets are created equal.

In this chapter you'll learn about televisions for home theater. TVs have become such a mainstay in our lives that we'll dispense with the obvious topics such as "What is television" and concentrate on TVs used in home theater systems — the best kinds to use and the best features to have.

If you already have a television set that you want to use with your home theater system, you can safely skip this chapter. But you may wish to read the section, "Connectors and Controls," to make sure the TV set you have is suitable for use in a home theater system. Also, see Chapter 8, "Room Design and Ergonomics," for suggestions on the best place to locate your television for maximum home theater enjoyment.

TYPES OF TELEVISION SETS

There are two main types of television sets suitable for home theater use, as shown in *Figure 3-1*.

- *Direct-view* sets use conventional pictures tubes, the standard in television sets for decades. They are called direct-view television sets because you view the picture directly on the screen.

••• INSTALLING HOME THEATER •••

- *Projection* sets use one of many image or light projection techniques to display the picture on a large screen. You view the picture by looking at the projected image on the screen. We'll talk more about the unique particulars of projection TVs later in this chapter.

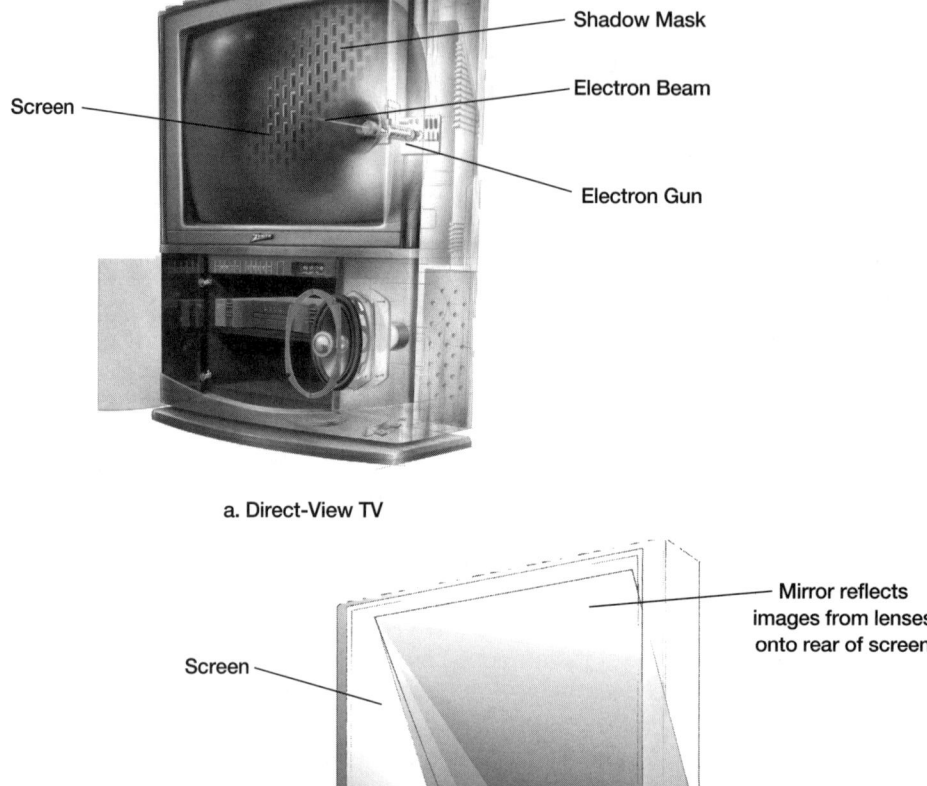

a. Direct-View TV

b. Rear Projection TV

Figure 3-1. Television sets for home theater come in two basic forms, direct-view and projection. Direct-view sets for home theater are traditional televisions with cathode-ray tube (CRT) screen sizes from about 25 to 40 inches. Projection models sport screen sizes of 40-inches and larger. Rear projection models, like the one illustrated here, reflect the picture off of a mirror and on to the back of the viewing screen. Front-projection models project the picture onto a reflective screen in a manner similar to a movie projector. (*Illustrations courtesy of Zenith Electronics Corporation.*)

3-2   The TV: Silver Screen of Home Theater

••• INSTALLING HOME THEATER •••

TV sets—direct-view or projection—can be further divided into three more groups: standard televisions, monitors, and TV/monitors. The difference is in the kind of signals that the TV can accept.
- Standard TVs accept only RF (radio frequency) signals received on VHF and UHF broadcast channels—channels 2 to 13 for VHF, and 14 to 69 for UHF.
- Conversely, monitors accept only baseband audio and video signals, such as those as provided by a VCR, video disc player, or A/V receiver. Baseband signals are the outputs of video sources that don't need tuning for the RF signals. Because they are not modulated on a radio frequency signal, quality provided by baseband video and audio is superior to that provided by an RF signal.
- TV/monitors offer a combination of input capabilities—they can receive both RF and baseband audio and video signals.

Throughout this book, we don't make a distinction between TVs, monitors, and TV/monitors. When we say "TV" we mean a TV/monitor, since this is by far the most common television set in use today. Most likely, it is the kind you already have—or will get—for your home theater system.

THE PREFERRED SIZE FOR HOME THEATER

TVs come in all shapes and sizes. The nature of home theater requires a set with a minimum screen size of 25 or 27 inches or larger for the greatest viewing enjoyment. Larger rooms that accommodate larger audiences demand larger screens, starting at 32 inches and zooming to 70 inches and beyond. (Remember: TV screens are measured diagonally, from one corner to the opposite corner.) "Large screen TVs" are generally defined as larger than 27 inches. These TVs can be direct-view or projection models.

Though there are no steadfast rules, the information presented in *Table 3-1* suggests television viewing screens based on the size of the room and the number of people in the audience (taking into consideration typical seating arrangements). Also provided in the table is the recommended screen-to-audience distance, based on the minimum screen size.

Room Size	Room Dimensions	Practical Maximum Audience Size	Recommended Screen-to-Audience Distance	Minimum Screen Size
Small	10' x 10'	3 persons	6 feet	25-inches
Medium	10' x 13'	4 persons	6 feet	27-inches
Large	12' x 15'	6 persons	7 feet	32-inches
Extra Large	14' x 17'	8 persons	8 feet	36-inches

Table 3-1. Room size and suggested TV screen size.

When selecting a screen size for your home theater system, you should take the following into consideration:
- The size of the room largely dictates the size of the TV. A small room is not suitable for use with a TV equipped with a very large screen. To minimize eye strain, the audience should not be seated closer than about three times the size of the screen. For example, if the screen measures 40 inches, the audience should be seated no closer than about 120 inches, or approximately 10 feet.

••• INSTALLING HOME THEATER •••

- Large screen TVs are physically large, and take up considerable space. The typical 40-inch projection TV, for example, takes up roughly the same floor space as a love seat. If the room is small, or already contains lots of furniture, you may have trouble fitting everything in.
- There is no rule that says large screen TVs are only for large rooms or large audiences. While you may be the only person watching your large screen TV, you might prefer its size because the image is bigger. However, the inverse is not true: a small TV will be dwarfed in a large room.

Connectors and Controls

For maximum picture quality, the TV set you use should be equipped with separate baseband audio and video input jacks that allow direct hookup to the audio and video outputs of a VCR, videodisc player, or A/V receiver. If your TV has only a standard "antenna-in" jack, you will find your connection options limited and picture quality impaired, and you may wish to upgrade to a newer set. Today, almost all TVs with screen sizes over 25-inches provide multiple inputs, including separate antenna and audio/video jacks. If replacing your set is not practical, there are two additional alternatives you may wish to consider:

- Use your VCR to convert the audio and video signals from the other components of the home theater system to channel 3 or 4 for reception on your television. See *Figure 3-2* for an example of how to connect your VCR in this manner. Note that this approach is not recommended as it reduces the flexibility of your system.
- Use a separate RF (radio frequency) modulator to convert the output of your A/V receiver to channel 3 or 4 for reception on your television.

Note: RF modulators, either built into VCR or stand-alone, typically do not process stereo sound (unless otherwise noted in the instructions for the product). Therefore even if your TV is stereo, you will not get stereo sound when the signal is processed by the RF modulator of a VCR.

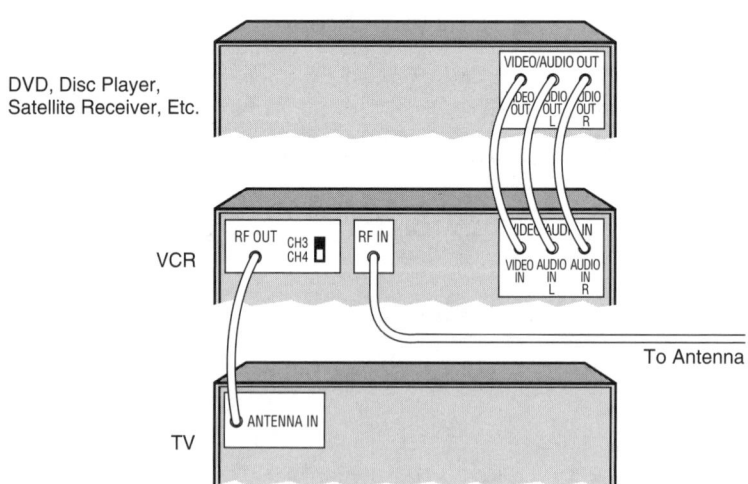

Figure 3-2. Use this hookup diagram if your TV lacks direct audio and video inputs. The VCR acts to convert the audio and video signal from your home theater components to channel 3 or 4 for reception on your TV.

3-4   The TV: Silver Screen of Home Theater

••• INSTALLING HOME THEATER •••

TVs WITH BUILT-IN SPEAKERS

The latest TVs generally come with improved audio and speaker systems. These systems are adequate, but usually don't measure up in home theater applications. You may be able to use the built-in speaker of the TV as the center channel speaker, and connections on the rear of the television may provide for this connection. Connect the center channel speaker terminal on the A/V receiver to this input jack on the TV. However, for best results, you should use a separate center-channel speaker.

STANDARD FEATURES

Following is a list of standard features found on most larger-screen television sets. None of these features are critical for home theater applications, but many help improve the picture and sound quality, or make the set easier to use.

- *Direct audio/video inputs* let you connect the TV directly to the audio and video out jacks of a VCR or A/V receiver. Many better televisions provide multiple inputs; you can select the input you want from a button on the TV or remote control.
- *Frequency-synthesized tuning* does away with the need for manual fine-tuning. Most modern sets are also "cable-ready," which means they can tune into the special cable channels as well as conventional over-the-air channels. For cable reception, the channel must not be scrambled; to view scrambled cable channels you still need a cable decoder box, available from your cable company.
- *On-screen displays* show you the channel number, and perhaps the time. Most television models also use on-screen displays to indicate brightness, contrast, color, tint, and other settings information. A few models let you add descriptive text to each channel, so when you change channels you know it's ABC, CBS, NBC, PBS, or some other station.
- *Unified remote controls* allow you to turn the set on and off, adjust the volume, and change the channel without getting out of your easy chair. Many remote controls also can operate other home theater equipment, such as a VCR or a DSS satellite receiver.
- *Closed captioning* lets you see the words on screen that are spoken by actors and narrators. You can turn the closed captioning off it you don't want to see it.
- *Better sound* with larger speakers and more powerful amplifiers. You can often use the speaker(s) built into the television as the center speaker for your home theater system—though a separate speaker is the preferred choice.
- *Multi-channel television sound (MTS)* lets you hear your favorite programs in stereo. (Not all cable companies provide stereo signals, however; check with yours to be sure.) Most sets also provide an SAP (Separate Audio Program) channel, usually reserved for a foreign language.

ADDITIONAL DELUXE FEATURES

If you're in the market for a new TV, you may want to look at these latest "deluxe" features found on some higher-priced models. As with the standard features described above, these added niceties are not required for home theater, but may help improve the picture and sound quality.

- *Video level presets* let you define one or more settings for watching TV under different circumstances. This compensates for the differences often found in signal

The TV: Silver Screen of Home Theater   3-5

quality between cable, VCR, DVD, and other program sources. If you're watching cable, for example, switch to the "cable" setting; if you're watching a tape through the VCR, switch to the "VCR" setting.
- *Flat picture tubes* reduce glare and distortion.
- *Color temperature adjustments* let you control the overall color output balance of the picture tube. The "industry standard" color temperature is 6,500 Kelvins (a Kelvin is a unit of temperature measurement), which tends to produce a "warm" or reddish picture. If you prefer a "cool" or "blueish" picture, you can increase the color temperate accordingly.
- *Digital circuits* can improve the picture and performance of the set, as well as provide new features. One handy digital feature is called PIP, or picture-in-picture. With PIP, you can watch two shows at once: the main show appears on most of the screen, and another show on another channel or on the VCR, for example, appears in a small box in the corner of the screen.
- *Multiple RF inputs* let you select the antenna or cable source you want to view. This feature is particularly handy when you have both cable TV and an outdoor antenna.
- *Built-in surround sound* gives you a home theater system without a separate A/V receiver with surround sound outputs. The television usually includes amplifiers for speakers (at least the center speaker and the rear-channel speakers). Though the set may include the surround sound circuitry, in most cases a separate A/V receiver will produce better sound and offer more listening flexibility.

Figure 3-3. Built-in television guide, such as StarSight™ or TV Guide Online™, provides on-screen display of TV listings. You can view programming by date and time, or by category. If you find something you like, you can select it, and the proper channel is automatically tuned for you.

- *Channel lock-out* lets you skip any channels you wish, such as adult-type programming. The viewer must enter a password on the remote control in order to view the locked-out channel.
- *High-definition video inputs* help improve picture quality. Some better TVs also offer a separate S-Video jack for connection to a VCR, DVD player, satellite dish, or A/V receiver that supports S-Video. A few very high-end models (mostly projec-

tion sets) offer component video inputs. These require a compatible component output from your VCR, disc player, or satellite dish. See Chapter 2, "Program Sources," as well as Chapter 5, "Hooking Up the Home Theater System," for more information about video devices and TVs that use S-Video and component video connections. Chapter 2 provides a background on how standard and alternative forms of video signals are used.

Understanding Comb Filters

If you've shopped for a television recently, especially a large-screen TV, you've probably seen a notation about the "comb filter" used in the set. A comb filter is a device that separates the main components of a video signal—specifically the color component (called *chrominance*) from the brightness component (called *luminance*). Once separated, these signals can be processed by the proper circuits inside the set. The result is an improved picture.

All TVs have a comb filter, but not all comb filters work the same way. Many modern sets use digital comb filters, with two types in common use: 2D and 3D adaptive. The 3D adaptive variety gets the nod of most television experts. High-quality digital comb filters are considered a "must" for large screen TVs, because of their screen size. Other TVs, particularly older models, use analog comb filters, which are not considered as good as the digital variety.

You can judge the quality of the comb filter in any given set simply by watching the picture. Look for something called "dot crawl"—which appears like tiny bubbles moving up or down the picture. Dot crawl is especially noticeable between two sharply defined vertical objects. The effect is most obvious when the objects are colored. An animated picture, like a Disney cartoon, is a good test for a comb filter.

Recall from Chapter 2 that some TVs and video components offer a variety of interconnections. A common interconnection for a TV is a *composite* video input, so called because it is a "composite" of the chroma, luminance, and synchronization that make up a television signal (composite video is another name for baseband video, which we spoke about earlier). Other input types include S-Video and component video:

- With S-Video, the comb filtering is done in the source device, not the TV. Therefore, the quality of the comb filter in the program source device (video disc player, for example) will determine the quality of the picture you see on screen. The comb filter in your TV is bypassed.
- With component video, comb filtering is not used because the chroma and luminance signals are not combined in the same way they are for composite video. Again, the comb filter in your TV is bypassed. Because the signal combining/separating process is avoided, the picture quality is generally the highest when using component video.

Of course, to enjoy the possible improvements with S-Video or component video, both your TV and your program sources must be equipped with compatible connections. If they are not, you may be able to use an after-market comb filter accessory. These connect between the video source device equipped with a composite output, and your TV equipped with an S-Video input. Your TV *must* have an S-Video input, or the separate comb filter accessory will not work.

Tips for Buying a New TV

Consider these buying tips if you're in the market for a new television set.
- View the picture under lighting conditions similar to your home. Some showrooms are unnaturally dark, making the picture on the TV screen look brighter. Turn on some lights, as you probably would at home, then view the picture.
- If the reception is poor you may get the wrong impression about an otherwise quality set. Consider bringing along one of your own video tapes and have the sales person play it. Viewing the tape before you go shopping will give you an idea of the minimum picture quality you wish to achieve for your home theater.

What About High-Definition Television?

The television broadcast system currently used in North America was devised in the 1940s and 50s to work with tube-based TVs receiving traditional analog RF signals. Since the invention of transistors, the development of integrated circuits and, especially, the advent of digital electronics, television broadcasting has become a terribly outdated technology. For more than a decade, manufacturers have developed a new television system that promises higher quality images. The manufacturing and broadcast standards for this system, called HDTV (high definition TV) or DTV (digital television), have now been finalized.

In the United States, the Federal Communications Commission has adopted a new digital broadcast standard and has awarded additional TV channel frequencies to stations so they can convert to digital broadcasting and implement this new technology across the country. By 2006, all "old style" analog broadcasts may be disallowed and high-definition digital signals will be the standard. The digital standards will provide clearer, sharper, and wider pictures (similar to pictures on a movie screen) and, using signal-compression techniques, more television channels will be on the air. (Several standard-definition digital channels can be compressed and sent in the same amount of frequency bandwidth as one analog channel.) In order to receive these digital channels you will need a DTV digital television.

So, with high-definition digital television "in our future" does it still makes sense to buy a traditional, analog television today? The answer is "yes." It will take years for television stations to switch over to digital transmission. And when they do, you'll be able to attach a digital-to-analog converter to your television (or VCR) in order to receive digital signals. (This will be similar to the technology used today to convert satellite signals received by small-dish DBS systems.) After 1998, some manufacturers will also offer combination analog/digital television sets that can receive both kinds of signals.

Televisions with Increased Aspect Ratios

MOVIES VS. TELEVISION

Sit down at a movie theater and you'll see the screen is much wider than it is tall. So-called "wide-screen" movies have been the mainstay of Hollywood since the 1950s when CinemaScope™ and similar technologies came into use. These days, movies are commonly released in either of two aspect ratios: 1.85:1 or 2.35.1. That is, the screen is 1.85 or 2.35 times wider than it is tall. The traditional TV set uses an aspect ratio of 4:3, or approximately 1.33:1. A relatively new class of TVs, mostly projection models, offer a physically wider screen, with an aspect ratio of 16:9, or about 1.78:1, which is very close to the 1.85:1 aspect ratio of many Hollywood motion pictures.

THE 16:9 ASPECT RATIO

The 16:9 aspect ratio of these television models is selectable. You can watch a regular 4:3 program it in standard mode, which displays an almost-square picture in the center of the screen. Dark gray or black bands are shown to the right or left where there is no picture. Or you can view the picture in "expanded" mode, which stretches the picture to fit the screen. Surprisingly, this does not produce a grossly distorted picture because the TV adds more stretch to the edges of the frame than the center. Actors in the center of the screen appear almost normal size.

The 16:9 aspect ratio comes in handy when viewing DVD and DSS programs delivered in wide-screen format, as well as "stretching" movies that are shown in standard *letterbox*. (Letterbox displays the picture in the same wide-screen form as seen in the movies, but on a regular 4:3 television screen black bars appear at the top and bottom of the screen.) *Figure 3-4* shows these different aspect ratios.

a. 4:3 Standard TV

b. 4:3 Letterbox Format

c. 16:9 Aspect Ratio

Figure 3-4. a.) Standard televisions display the picture on a screen with an almost-square 4:3 aspect ratio. b.) Letterbox formatting, with black bars at the top and bottom of the picture, is used to show a wide-screen picture on a standard 4:3 TV screen. c.) A new class of TVs use a 16:9 wide-screen aspect ratio to display the picture. The 16:9 aspect ratio will be the standard for HDTV digital sets. These newer sets also can display the traditional 4:3 aspect ratio pictures, either "stretched" to fill the screen, or with black bars on either side. (*Illustrations courtesy of Zenith Electronics Corporation.*)

SWITCHING ASPECT RATIOS

Most DVD disc players and DSS satellite receivers allow you to switch aspect ratios of some program material. The aspect ratio of a picture provided by DVD or DSS is a different issue than the aspect ratio of a television set. For a TV, the aspect ratio is determined by the physical measurements of the picture screen. For DVD or DSS, the aspect ratio is controlled by software; you see the result on any type of television, either 4:3 or 16:9 sets.

For example, when viewing many movies recorded on DVD, as shown in *Figure 3-4*, you can use the remote control to select between "normal" (4:3), or "wide-screen" (16:9) aspect ratios. When viewed on a normal television, a 4:3 picture will fill the screen, but a 16:9 picture will appear with the traditional letterbox bars at the top and bottom, as shown in *Figure 3-4b*.

Conversely, when viewed on a wide-screen television, a 4:3 picture will be centered in the middle, with gray or black bars to the sides of the screen (or, you can choose to "stretch out" the 4:3 image to fill the whole picture). A 16:9 picture will properly fill most or all of the screen, as the movie did when projected in theaters.

Wide-screen TVs can even automatically switch between 4:3 and 16:9 modes when you change the aspect ratio of the program played on the DVD or DSS unit. The TV detects the aspect ratio being displayed, and assumes that you want to watch a 4:3 picture in 4:3 mode, and a 16:9 picture in 16:9 mode. For the automatic detection to work, the TV and the DVD or DSS unit must be equipped with component video connections.

LARGE SCREEN TVS

Large screen, direct-view TVs are available up to about 40 inches in size. For larger screens, projection TVs are the answer.

DIRECT VIEW

Direct-view picture tube models are regular television sets, except the picture tube is physically larger than normal. Today, direct-view, large screen TVs are limited to about 40 inches, with 32- and 36-inch sets being very popular. For the most part, direct-view models offer the brightest, sharpest pictures of all types of large screen TVs. Prices can be steep, and the sets are heavy, large, and fragile.

REAR PROJECTION

There are two types of projection TVs, as shown in *Figure 3-5*. Rear-projection models are considered the "standard" in large screen TVs. The set is composed of a high-intensity projection unit housed in a cabinet. The projector beams a color television image onto a set of mirrors that reflect the image onto the back of a translucent plastic screen. The screen typically uses thin horizontal "grooves" to increase brightness when viewing the picture from either side. This screen can be easily marred and scratched, so many sets use a separate plastic covering to prevent damage. One disadvantage of rear-projection models: the screen is somewhat reflective and, for maximum picture brightness, the TV should be viewed in a dimly-lit room. Sizes of rear-projection models extend to 72 inches and beyond, with 40- and 50- inch models being the most popular.

••• INSTALLING HOME THEATER •••

From time to time, the projection unit in rear-screen televisions must be re-aligned, so that they continue to produce clear, sharp pictures. Most newer, rear-projection models come with an alignment or "convergence" test mode; place the TV in test mode, and adjust the convergence of the test pattern.

a. Direct View TV

b. Rear-Projection TV

c. Digital HDTV receiver/decoder (left) and Front Projection Monitor

Figure 3-5. Large screen TVs are available in a.) direct view, b.) rear-projection, and c.) front projection models. Direct view sets offer the sharpest pictures, but screen size is limited to 40 inches. Rear and front projection sets can display pictures of 70-inches or larger. (*Illustrations b and c courtesy of Zenith Electronics Corporation.*)

FRONT PROJECTION

Popular among the *cognoscenti* of home theater are the two-piece, front-projection televisions, which have separate projection units and screens. The projection unit is often suspended from the ceiling, usually behind the audience. If ceiling mounting is not possible, it can be placed in a cabinet, and located in front of the audience.

The TV: Silver Screen of Home Theater   3-11

There are two main types of projectors: cathode ray tube (CRT) and liquid-crystal display (LCD). In the former, a set of three CRTs are used to produce the three primary television colors—red, blue, and green. The light from these cathode ray tubes is focused by lenses onto the screen. In the LCD variety, a single, bright white light source is placed behind an LCD panel. Light shines through the panel and projects the picture onto the screen.

The screen for a two-piece front projection setup can be permanently attached to a wall, or it can be rolled down for viewing. Wall-mounting allows the screens to be curved, which increases brightness for those sitting in front of the screen. The screen material is highly reflective, but is not always stark white. A silver fabric-like screen often is used.

Of the two types of front projection televisions, the CRT units tend to be the most difficult to set up. Proper convergence alignment is critical to ensure that all the colors blend together properly on the screen. Periodically the projection unit must be re-aligned, which may require the aid of a knowledgeable service technician. Re-alignment is also required if the TV is moved.

Shopping for a Large Screen Projection TV

Room lighting, viewing distance, and viewing angle all play important roles when judging a large screen television, particularly projection models. You may be unfairly biased if you "test drive" a large screen TV under improper lighting conditions, when sitting too close or too far from the screen, or when viewing the screen at angles greater than about 30 degrees. Big screen TVs, especially projection units, require a more-controlled viewing environment. You should strive to achieve that environment both when you are looking for a set to buy and, more importantly, when watching it at home.

That said, you should also test the model(s) you like best by turning on the lights and viewing from extreme angles. The better large screen televisions are more tolerant of variations in the "ideal" viewing environment. You'll probably want to stay away from the projection model that looks terrific straight on, but loses almost all of its brightness and clarity when viewed at any other angle. A limited viewing angle is particularly troublesome for large families, who may be seated all around the home theater room.

ALLOWABLE SPACE

When you choose a projection TV, measure the amount of space available in the room where the TV will be placed. Choose a spot away from windows and doors. Sunlight streaming through a window or open door not only can overpower and wash out the image but also can damage the glass or plastic material used for the screen.

BRIGHTNESS

Manufacturers flaunt lots of impressive facts and figures about their projection units, and some of the terminology may be a bit unfamiliar. For example, screen brightness on many projection sets may be specified in "foot lamberts," a measurement of luminance or brightness within a single square-foot area. Most manufacturers measure the brightness on the screen when the TV is displaying a totally white picture, called "peak white." Some manufacturers list the screen brightness for an "average" picture with both light and dark areas. Unfortunately, they don't often tell you which standard they are using—peak or average—so always use the screen brightness specifications only as a guide. The higher the foot lambert value (80 to 100 is a good starting point) the better. Remember: the larger the screen, the less intense the image. In the past, manufacturers have improved brightness by increasing the voltage applied to the picture tubes inside the projector, resulting in shorter tube life. When choosing a projection TV, ask the dealer how long the tubes should last. Anything less than four or five years indicates that it might be an expensive TV to own.

LENSES

The lenses employed in projection sets also have their own terminology, and there is much controversy over which kind delivers the best picture. Most lenses used in projection TVs are large—about four or five inches in diameter. You might read that one projection TV uses "five-element, f-1.0" lenses. The number of elements in a lens simply tells you how many groups of glass (or plastic) are used. Usually, but not always, the more elements, the better the lens. As in photography, the "f" number is the "speed" of the lens. The lower the "f" number the more light that passes through the lens resulting in a brighter picture.

Many projection TVs have plastic lenses. Even though they are plastic, they may be of excellent optical quality, rivaling many glass lenses of the same size. Typically, glass lenses have a higher "f" number, which means that less light passes through. Although glass lenses typically mean a sharper picture overall, the faster plastic lenses can pass up to 30 percent more light—another tradeoff. A good projection TV might have five-element f-1.0 lenses, but the best test of brightness and picture quality is the one that involves your own eyes. Just sit down in front of the screen and watch. Does the image seem sharp and in focus, or is the picture dull and lacking in contrast? Are the colors separating? Does the projector deliver a picture you could watch for hours on end?

SUMMARY

The TV is the "silver screen" of your home theater system. If you're buying a new set for your home theater, choose the type, size, and features of the TV carefully. You'll save money and get a better TV. Or, if you already have a television suitable for your home theater, make sure it has the proper video connections so you can enjoy the sharpest picture possible. In this chapter you learned about the "visual" side of your home theater. In the next chapter you'll learn about the sound component of home theater, including the A/V receiver used to tie all the components together.

Sound – The Other Half of Home Theater

Alexander Graham Bell was the first person to experiment with "stereophonic" sound produced by means of an electronic device. As described in the July, 1880 issue of *The American Journal of Otology,* Bell used four of his newly-developed telephones. Two telephones, spaced a few inches apart, served as microphones. Bell, located in another room, listened with the earpieces of the other two phones.

Here's what Bell found: even though he wasn't in the same room as the sound source, his brain accurately reconstructed the arrangement of the microphones—he felt he was actually there in the room. Bell used microphones that were placed close together, about the same distance as ears positioned on either side of a human head. The "stereo" effect, Bell discovered, could be expanded by increasing the distance between the microphones. Increasing the distance between the microphones appeared to increase the size of the listener; decreasing the distance appeared to decrease the size of the listener.

In this chapter, you'll discover how sound contributes to the home theater environment. You'll read about the all-important A/V (audio/video) receiver, Dolby Stereo, and the use and placement of speakers in your home theater den.

The Science of Stereo

Today, stereo is so common we take it for granted. And the stereo effect—creating and altering the listener's sound environment with multi-channel sound—forms the backbone of home theater. Without stereo, you won't enjoy the full effect of being surrounded by sound.

The stereo effect is made possible by a number of factors:
- Loudness differences: sound that is louder from one speaker than another appears to be closer to you.
- Quality differences: raspy or weak sound appears to come from a distance, or through an object.
- Arrival-time (phase) differences: sound that arrives in one ear before the other appears to be coming from a certain direction.

Our ears and brain also use the phase differences to determine the source of sound. Sounds in-phase (both ears hear the same sound simultaneously) will appear local, and straight ahead (or behind us). When our ears detect sounds that are out of phase (our ears hear the same sound but the sound waves are not in sync), our brain correlates that to a direction of the sound source. The typical stereophonic sound system uses two speakers positioned in front of the listener and placed a certain distance apart—the ideal separation is determined by the distance between the listener and the speakers. For ideal stereophonic reproduction, the speakers should be placed the same distance apart as they are from the listener, as shown in *Figure 4-1*.

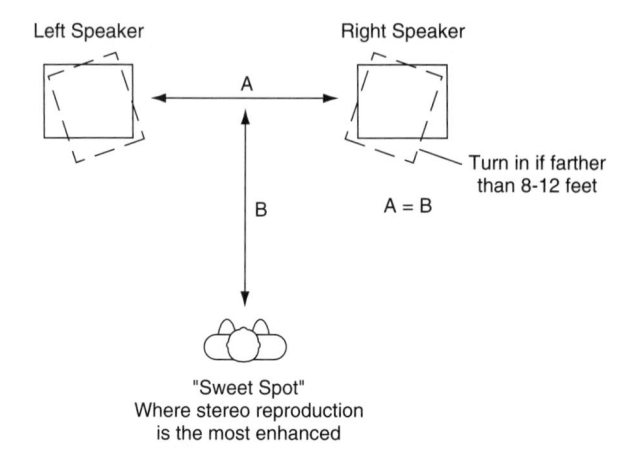

Figure 4-1. Ideal stereophonic reproduction is achieved when the two speakers are placed at about the same distance as the speakers are from the listener. The speakers may be turned inward if the distance is greater than about 8 to 12 feet. This creates a "sweet spot" at the location of the listener.

One of the problems of the two-speaker arrangement is that, as the listening distance increases, a "sonic hole" can appear between the two speakers. Paul W. Klipsch, a pioneer in the hi-fi industry, was one of the first to add a center speaker to a stereo in order to "fill in" sonic holes caused by distant right- and left-front speakers. Though Klipsch first introduced the center speaker idea decades ago, it didn't really catch on with the general public until the advent of home theater.

Three speakers in front of the audience are still not adequate to fully surround a listener with sound. Try this experiment: Sit in front of traditional hi-fi speakers. Turn on some music and close your eyes. More than likely, you will get the distinct impression the music is coming from some point directly in front of you.

When you're watching a movie you want to be immersed in the action, feeling as if you are in the middle of it, not watching as a spectator from some distance. Speakers placed behind you provide "surround" or ambient sound sources. These speakers don't need to be as loud as the front speakers. They merely serve to reinforce the sound coming from the speakers in the front.

Recall that one of the ways to create a stereo effect is to change the arrival time of sound. By delaying the sound between the front and rear speakers, it is possible to artificially expand the listening area. Of course, the room stays the same, but our brain interprets the sound as coming from a much larger area. For this reason the surround (also called rear) speakers are a key ingredient in any home theater system. Without them you will never feel you are part of the on-screen action.

••• INSTALLING HOME THEATER •••

THE A/V RECEIVER

The A/V receiver is the nerve center of the home theater system, performing a number of critical functions. Among the most important:

- It switches among different video and audio inputs. Most home theaters have multiple program sources, such as VCR, video disc player, and broadcast (via antenna, cable, or satellite receiver). You use the A/V receiver to select the program source you wish to watch.
- The A/V receiver decodes the Dolby Surround sound track information. Most modern movies—on video tape, video disc, satellite dish, and even through the cable and off-the-air—are presented in Dolby stereo. As you'll read later in this chapter, Dolby stereo expands the normal two-channel/two-speaker stereo to four-channel stereo that is heard through five or even six speakers.
- It serves as the stereo receiver, amplifying each sound channel and powering each of the speakers in the home theater system.

Ancillary features of most A/V receivers also include the ability to tune in AM and FM radio broadcasts, and amplify sound from music CDs, cassette tape decks and phonographs.

ADDITIONAL SOUND MODES

In addition to reproducing Dolby stereo, home theater A/V receivers typically provide additional sound modes, most of which are synthesized from standard two-channel stereo. For example, most A/V receivers have a "concert hall" mode that uses the surround speakers in the back of the room to expand the apparent listening area, making it seem as if you're in a concert hall. More sophisticated A/V receivers offer additional sound modes that employ some type of pre-programmed echo and tonal equalization to create the sound ambiance of an intimate nightclub, a stadium, or a movie theater, simply with the press of a button.

CHANNELS AND WATTAGES

A/V receivers are classified by both the number of sound channels they provide and the power, expressed in wattage, delivered to the speakers. The wattage of an A/V receiver may be expressed in total watts for all sound channels, or for each sound channel. For example, a given A/V receiver may produce 400 watts. Or it might be "rated" at 100 watts for each of four separate sound channels: left-front, right-front, center-front, and surround speakers.

Whether you select an A/V receiver based on overall wattage or watts per channel, it is important that the receiver provide adequate power to each speaker. If not, the sound quality of your home theater system may be compromised. As a general rule, you need an A/V receiver capable of delivering a minimum of 30 to 50 watts per channel. If the wattage for the A/V receiver is given as a sum of all channels, the unit should provide not less than about 120 total watts.

If you use a subwoofer in your home theater system (see "Speakers Used in Home Theater," later in this chapter), your A/V receiver should be capable of producing not less than 150 to 200 watts total, with 80 to 100 watts (total) allocated to the front speakers alone.

Sound—The Other Half of Home Theater   4-3

DISCRETE CHANNEL INPUTS

A/V receivers are designed to let you switch among various program sources, including a VCR, video disc player, or satellite receiver. Press a button on the front of the receiver or on the remote control and you select the program source you wish to view. Both the video and audio signals are selected simultaneously, so that you hear the sound that accompanies each program source.

All A/V receivers provide separate stereo (right and left) inputs from various program sources. A few models also provide discrete channel inputs—that is, separate inputs for each speaker connected to the A/V receiver: right-front, left-front, center-front, and surround (rear) speakers. In the case of A/V receivers that are "Dolby Digital" (described later in this chapter), six channel inputs are provided. The two additional channels provide stereo signals to separate right- and left- surround speakers, and power a separate subwoofer.

DOLBY SURROUND

Dolby is a name long associated with sound systems. During the 1960s, inventor Ray Dolby developed a method of decreasing noise for audio taped recordings. This method—now universally known as Dolby Noise Reduction™—is still in use today.

There are other forms of Dolby sound technology in use today, including Dolby Surround and Dolby Digital. Though these systems are different in the way they work, they are both used for the same thing—adding more sound channels to a stereo presentation. Instead of having just two front speakers for the left and right stereo channels, Dolby systems have three or four additional speakers located at specific points in the room. The overall effect is to "surround" the audience with 360 degrees of sound.

DOLBY PRO LOGIC

Dolby Surround is a somewhat generic term used to denote several types of analog, multi-channel sound reproduction systems. The older "matrix" surround system was created in the 1970s for movie theaters. This system decoded up to four additional channels recorded on traditional stereo soundtracks. A main component of the system was two speakers in the back or side of the audience—called surround speakers—that provided ambient sound.

The matrix surround system was later improved with the invention of the Dolby Pro Logic system, which uses intelligent circuitry (such as DSP, or digital signal processing) to "steer" each channel of sound to its appropriate speaker. As with matrix surround sound, Dolby Pro Logic was introduced first in movie theaters, then began to appear in consumer video and audio components for home theater. Today, nearly all home theater components use the Dolby Pro Logic system.

There are five speakers in the typical Dolby Pro Logic system, with an optional sixth subwoofer.

- Two main speakers are placed in the left- and right-front area of the room. These are the stereo *left- and right-channel* speakers.
- An extra speaker is placed near the television screen. This is the *center-channel speaker*. It carries a mixture of the content of both the left and right stereo speakers (such as dialog, music, and main sound effects). This helps "center" the sound and make it seem that the dialog is coming from the TV screen.

••• INSTALLING HOME THEATER •••

- Two *surround speakers* are placed at the rear or side of the audience. These speakers carry ambient sound, as well as portions of the sound track. Typically, the sound from surround speakers is delayed a few fractions of a second from the sound coming from the front speakers. This has the effect of increasing the apparent size of the listening area—suddenly, your home theater den is the size of an orchestra hall! Note that with Dolby Pro Logic the audio from the two surround speakers usually is monaural.
- An optional *subwoofer* provides deep bass sounds, enhancing sound effects.

Note: With Dolby Surround the signal for the subwoofer is derived from the main left and right stereo speakers. Neither system of Dolby Surround—matrix or Pro Logic—provide a separate channel for the subwoofer. However, as you'll discover in the next section, the Dolby Digital system does offer a separate channel for the subwoofer.

One of the benefits of the analog Dolby Surround systems—and why they have become so popular—is that they are compatible with monophonic and standard stereo systems as well. You can play a movie recorded in Dolby Surround on any VCR, for instance. If you have a stereo VCR connected to a stereo sound system, you will hear the sound track in stereo. If you have a stereo VCR connected to an A/V receiver equipped with Dolby Surround circuitry, you will hear the sound track in full Dolby Surround, complete with the additional sound channels.

DOLBY DIGITAL

Dolby Pro Logic systems provide an excellent "sound stage" for most any home theater. Sound appears to "magically" emanate from different speakers and you get the very real sense that you are part of the action on the TV screen. Yet, the Pro Logic system has one inherent flaw—it is based on analog encoding and decoding schemes that may not always accurately "steer" each sound to the proper speaker. The Dolby Digital system is an improvement over Dolby Pro Logic. As with Dolby Surround, Dolby Digital started life in movie theaters. With Dolby Digital (sometimes referred to as AC-3), the sound channels are never mixed. Rather, they are kept separate, or discrete, and each channel delivers its sound to the correct speaker and to no other.

Currently, Dolby Digital is more expensive than Dolby Pro Logic. You'll see movies presented in Dolby Digital available only in the two laser video disc formats, as well as some digital satellite systems. VHS tapes cannot be encoded with Dolby Digital. Of course, movies with Dolby Digital sound tracks always provide a regular sound track as well and can be played on Pro Logic equipment.

On a video disc, the two sound tracks are stored with the video signal. The standard stereo sound track is available for those video disc players and A/V systems that lack Dolby Digital capability. Note that all DVD disc players are capable of processing Dolby Digital sound tracks. Some DVD players provide their own Dolby Digital processors and offer six connectors on the back—one for each of the sound channels supplied by the system—for hooking up to a compatible A/V receiver. Other DVD players don't have a built-in Dolby Digital processor and must be used with an A/V receiver equipped with a Dolby Digital input or external Dolby Digital processor.

The speaker arrangement for Dolby Digital is similar to that for Dolby Surround (see *Figure 4-2*). One main difference is that the two surround speakers in Dolby Digital are stereo, not mono as they are with Dolby Surround. Also, the sound for the subwoofer is carried on its own channel; not derived from the two main front speakers. As such, Dolby Digital supports six separate (discrete) channels of sound. You will often see Dolby Digital described as "5.1," which means the number of channels in the system. The ".1" is the subwoofer. The bandwidth of the subwoofer sound channel is very limited (specifically, from 3 Hz to 120 Hz), since it only has to reproduce a narrow range of very low frequencies.

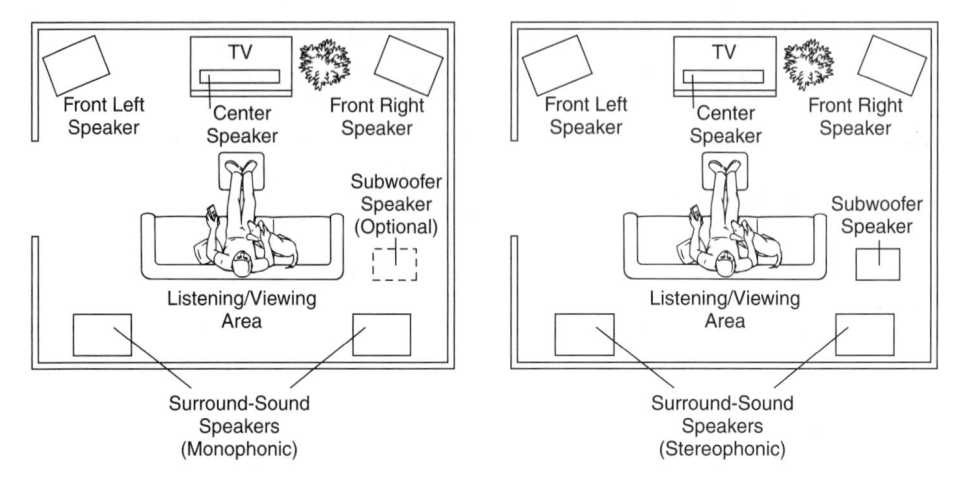

Figure 4-2. Dolby Pro-Logic and Dolby Digital use a similar speaker arrangement. With Dolby Digital, the surround speakers are stereophonic and a separate sound channel is reserved for the subwoofer. The subwoofer is optional, but recommended, with either Dolby Pro Logic or Dolby Digital.

It is important to note that, while Dolby Digital is capable of supporting six (or 5.1!) sound channels, not all programs provided in Dolby Digital format have all channels working. A movie can be presented in Dolby Digital but still be stereo, or even monaural, because some older films were never recorded with the extra sound channels. Therefore, even if the program is presented in Dolby Digital you will not enjoy full surround sound.

PROCESSING DOLBY DIGITAL

The Dolby Digital signal can be processed in any of three ways:

1. By an individual home theater component, such as a DVD player. The Dolby Digital signal is decoded and processed within the player. Six sound outputs are provided at the back of the player for connection to a "Dolby Digital ready" A/V receiver that supports individual inputs for each channel.
2. By an external decoder placed between the home theater component and a Dolby Digital-ready A/V receiver with individual channel inputs. The program source—DVD player, satellite dish, etc.—provides the raw Dolby Digital signal to the external decoder, which processes the signal into the six sound channels.
3. By an A/V receiver equipped with its own Dolby Digital processor. The receiver accepts the raw Dolby Digital signal, decodes it, then routes the sound to each of the six speakers. No external decoder is required.

Of these approaches, the second and third are generally considered better because they can decode Dolby Digital sound tracks regardless of the program source.

Home DTS and THX

There are several additional sound systems in use in movie theaters, and like Dolby Surround and Dolby Digital, these sound systems have found their way into some home theaters. One is THX®, developed by Lucasfilm, headed by film maker George Lucas of *Star Wars* fame. THX isn't really a unique sound system in and of itself, but rather a collection of standards designed to ensure high quality reproduction of the sound track.

Like Dolby Digital, Home THX is based on six discrete channels of sound (actually, 5.1—five regular speakers, and one subwoofer). THX specifies dipole speakers—speakers with sound ports on the front and back—for the surround speakers, as well as higher dynamic range subwoofers and enhanced power output of the amplifier to ensure adequate sound reproduction. An A/V receiver that has been "THX certified" has been found to follow the strict criteria for sound reproduction. By their nature, all laser disc players with Dolby Digital outputs are capable of playing any THX certified program material (you will always see the THX logo on the disk jacket); you do not need to buy a special THX disc player to view a THX certified disc.

Another sound system gaining popularity in home theater systems is DTS®, developed by the company of the same name. Like Dolby, DTS began as a movie theater sound system. At the theater, the DTS soundtrack is stored on a CD, played in synchronization with the movie. DTS for home theater is available on selected, pre-recorded video discs labeled with the DTS logo. In a home theater setup, the soundtrack signal is encoded with the video on a laser disc; there is no separate soundtrack CD in the home DTS setup.

Speakers Used in Home Theater

Speakers and their placement in the room play a critical role in producing quality sound for your home theater system. Inexpensive or improperly placed speakers can actually detract from the enjoyment you might otherwise get from your home theater system, so it's important to consider them carefully.

LEFT- AND RIGHT-FRONT SPEAKERS

A lot of attention is given to the left- and right-front speakers. However, they are no more important than the other speakers in your home theater system. Many home theater enthusiasts use their existing stereo hi-fi speakers for the left- and right-front speakers. This is perfectly acceptable as long as the speakers are still in good condition. Speakers do "wear out" over time; specifically the cone of the speaker weakens with age and use. You may not even be aware of this deterioration until you listen to your home theater system through new speakers. If your speakers are getting old, or you don't already have a set, look for a pair of speakers that can handle the wattage of your home theater A/V receiver and that integrate well with your decor.

Left- and right-front speakers are placed on either side of the television screen, and located the same distance apart as they are from the listener. You can use "tower" or compact bookshelf speakers, depending on your preference and the layout of the room. You *do not* need big speakers for "big home theater sound." Home theaters derive their unique *sound stage* through the use of multiple speakers, not individual speakers that attempt to overpower the listener.

••• INSTALLING HOME THEATER •••

CENTER-FRONT SPEAKER

Most home theater systems use a single speaker placed near the TV. This speaker "fills in" the sound to make it appear as though it's coming from the screen. Dialog and most on-screen sound effects are heard through the center-front speaker.

Place the speaker on top of the TV, as shown in *Figure 4-3,* or on a shelf above the TV. Avoid placing the center channel speaker below the TV, as the proximity to the floor often artificially increases the bass response of the speaker and the sound comes out "boomy" and distorted. The center-channel speaker needs to by physically close to the TV so that the dialog and other sound appears to come from the TV screen.

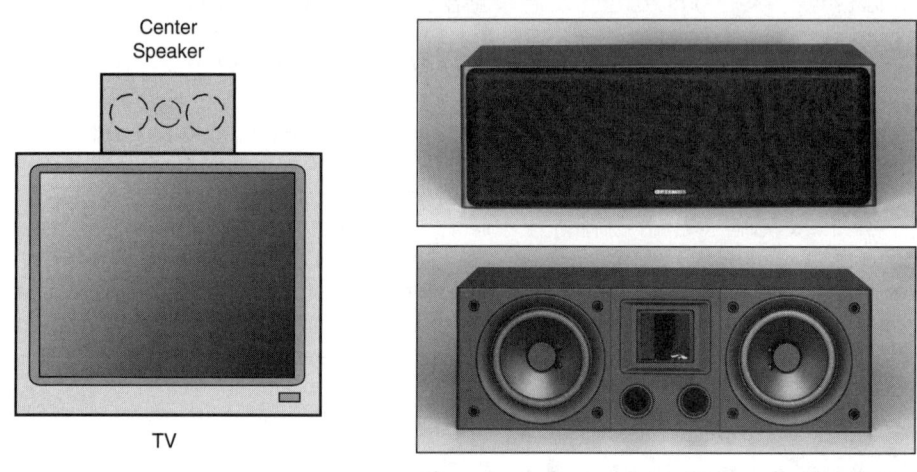

Figure 4-3. You can place the center front speaker (like the one shown on the right, with and without its grill) directly on top of the TV, or on a shelf above the TV. Don't worry about the speaker being too heavy for the TV; it is a rare television set that cannot bear the 5-10 pound weight of the typical front speaker.

Be sure to use only a magnetically-shielded speaker. This prevents the magnetic field produced by the speaker magnet from causing interference on the CRT television screen. If you notice a distorted picture, or a "rainbow" pattern in the picture close to the speaker, it indicates a poorly-shielded speaker enclosure. Replace the speaker with a properly-shielded model. Shielded speakers suitable for use as a center channel speaker in home theater systems are available at Radio Shack and other home entertainment stores.

Many television sets have their own internal speaker and, in a pinch, this speaker can be used as the center-front speaker. However, even the built-in speakers in the best quality TVs usually are inferior to even a budget-priced center-front speaker. You are always better off using a separate center-front speaker.

SURROUND (REAR) SPEAKERS

The surround or rear speakers are placed to the rear or side of the audience. They add ambiance and help provide the "all-encompassing" feel of home theater sound. Because surround speakers provide only ambient sound and some music, their size and design is not as critical as the three front speakers. Many home theater enthusiasts choose small, compact *bookshelf* or *cube array* units for the surround speakers.

••• INSTALLING HOME THEATER •••

These can be mounted on a wall, placed on bookshelves, or attached to special stands. Some surround speakers are designed to be mounted inside the wall to be as unobtrusive as possible.

If you're looking for a top-of-the-line home theater system, you may want to consider getting a pair of *dipole* surround speakers, like those shown in *Figure 4-4.* Dipole speakers produce sound through ports on both the front and rear of the enclosure. This special design helps increase the dispersion of sound from the surround speakers.

Figure 4-4. Small, bookshelf-size dipole speakers, shown with and without grills, are an excellent choice for the surround (rear) speakers of your home theater sound stage.

For best results, place the surround speakers at or slightly above ear level of the audience. Place the surround speakers no closer than about three feet from anyone in the audience to prevent the sound from these speakers from distracting the viewer. Though often referred to as "rear" speakers, surround speakers do not need to be located behind the audience. In fact, the current trend is to place the surround speakers to the side of the audience.

SUBWOOFER

The subwoofer produces low-frequency sounds, such as the rumble of an earthquake or the shaking of a jet passing nearby. Subwoofers, like those shown in *Figure 4-5,* provide more of a *sensation* than actual sound, though if you placed your ear near a subwoofer *(not recommended!),* you will probably hear some low-frequency sounds. Home theater systems typically use only one subwoofer because the origin of low-frequency sound is very difficult to discern. The very deep bass sounds from a subwoofer will appear to be coming from "everywhere," even when the subwoofer is placed in front of, behind, or to the side of the audience.

By their nature, subwoofer speakers tend to be large. The typical subwoofer needs a lot of power from the A/V receiver in order to produce sufficient sound levels. If you want to add an unpowered subwoofer to a home theater system, you must be sure your A/V receiver is capable of providing enough wattage. As a general rule of thumb, the A/V receiver should be able to deliver not less than 50 watts for each of the left- and right-front channels, and preferably 75 to 100 watts.

••• INSTALLING HOME THEATER •••

Should your A/V receiver not be capable of producing sufficient wattage, an alternative is to invest in a "powered" subwoofer. These speakers contain their own built-in amplifier. This arrangement frees your A/V receiver from having to provide the extra wattage to drive the subwoofer. The main disadvantage to powered subwoofers—apart from higher cost—is the need to locate them near a power outlet.

Figure 4-5. You can use a powered or unpowered subwoofer to add deep bass sound to highlight the sound effects from your home theater sound system. The subwoofer on the left, without its grill, is "front firing," while the one on the right is "top firing." Selecting the correct subwoofer depends upon where it will be placed in the room.

CHAPTER SUMMARY

This chapter detailed the sound portion of home theater. It discussed the use and main features of A/V receivers, Dolby Stereo, and speakers. Next, you will learn how to connect all these components together—along with VCRs, video disc players, satellite receivers, and other program sources—to create a functional home theater.

Hooking Up the Home Theater System

The moment has come! You've just purchased your new home theater system and are about to install it in your living room or family room. But where do all the components go? Which inputs and outputs should you use? Is there a problem with that 50-foot length of thin speaker wire? And are there any special considerations if you connect extra speakers to the system?

This chapter provides details on how to connect together all the components of a typical home theater system. It offers help on where to place the components; detailed illustrations for connecting the pieces together; discusses the right and wrong way to attach speakers; explains how to make your own coaxial cables; and more.

First: The Difference Between Wire and Cable

Throughout this book, we use "wire" as a generic term. A wire basically is a length of metal conductor surrounded by plastic or rubber insulation. For the most part, *wires* aren't used much in home theater systems, except to connect the A/V receiver to the speakers. Speaker wires are easy to spot as they use large metal conductors. The size of the conductor is necessary to assure maximum power transfer from the amplifier to the speakers. *Cables,* which come equipped with the proper connector for easily attaching the components of the home theater system, are used for everything else.

A cable is composed of one or more individual wires, fused or sheathed together. Most often, the cable is terminated using a unique connector. The type of connector usually (but not always) is determined by the application. A connector for a speaker wire is different than a connector used for hooking-up a VCR to an A/V receiver.

Cable, as shown in *Figure 5.1,* also generally denotes one or more wires protected by a metal shield that surrounds the wires. The shield reduces interference cause by outside electrical fields. The metal shield is connected to the chassis ground of the home theater system. With the exception of the speaker wires that connect the power amplifier and speakers, all cabling in your system should be made of the shielded type.

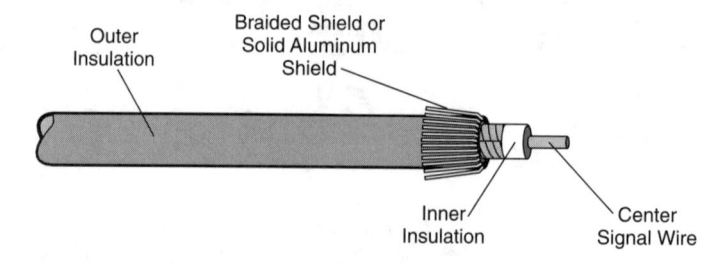

Figure 5-1. Shielded cable is composed of a center conductor, surrounded by an insulating dielectric. Braided wire forms a "shield" around the dielectric, and reduces interference from external sources.

COAX OR SHIELDED

Cables for home theater come in two basic types: coaxial and shielded. They are designed for different applications:
- Coaxial (or just "coax") cable is used to carry radio frequency (RF) signals, such as those broadcast over TV channels, from your antenna to your TV set or VCR.
- Shielded cable is used for baseband audio and video, the kind of signals available at the Audio Out and Video Out jacks of your VCR or A/V receiver.

Though the fundamental construction of the two types of cables is somewhat similar, coax is much thicker and bulkier than shielded. Shielded cable is the type most often used in home stereo systems to connect turntables, compact disc players, and tape decks to an amplifier, and is the type most familiar to hi-fi enthusiasts.

For a quality home theater setup, you should use coaxial cable only for connecting to the cable TV service and your off-the-air TV antenna. The reason: television signals are RF signals that are modulated to higher frequencies for ease of transmission and then demodulated to lower RF frequencies. Signal quality is reduced by this modulation. Thus, in order to maintain the highest quality video and sound, all components of your system should be inter-connected using shielded cable through the AUDIO and VIDEO connections on your home theater components. This will allow your system to use the unmodulated baseband signal to display the video and reproduce the audio.

CONNECTOR QUALITY

The connectors used on most "economy" video and audio cables are mass produced using copper or tin and then plated with nickel or chromium. The plating resists, but is not impervious to, corrosion and oxidation. Nickel is a marginal conductor of electricity, but its low cost and ease of manufacture make it the best all-around choice for economy consumer electronics products.

Better quality connectors use gold plating. Gold is perhaps the best conductor of electricity and, more importantly, it never corrodes. The downside of gold is that it is much more expensive than other plating metals. Gold-plated connectors cost two to four times as much as their nickel-plated cousins. You can purchase manufactured cables with gold-plated connectors already attached, or you can make your own custom-made cables by purchasing gold-plated connectors and high-quality wire. The cables and connectors are available at Radio Shack and elsewhere, in both the RCA phono and F-connector varieties.

Placement of Home Theater Components

You've purchased a new home theater system and the boxes are stacked in the center of the room. What now? The first order of business is to consider where you will place each component. Begin with the TV, since that will be the focal-point of the system. The TV should be located along a wall with the audience facing it. In most cases, you'll want to place the TV along the wall that has the cable or antenna outlet. This will allow you to attach the cable and antenna to the home theater components with the greatest ease.

However, you should weigh other factors before locating the TV at that spot. Just because a given wall has the cable or antenna outlet doesn't mean it's the best location. Ideally, there should be enough space to locate the other home theater components along this wall, including the right- and left-front speakers. The wall also should be positioned away from bright light sources; otherwise, you may experience excessive glare when trying to watch the TV.

Should you need to locate the TV on another wall, consider ways to extend or relocate the cable and antenna outlets to the TV and other home theater components. The same consideration applies to the electrical outlets. If the ideal wall in the room lacks an electrical outlet, you will need to consider ways to run a heavy-duty extension cord from an existing outlet, or have a new outlet installed in this wall.

Once you have identified the best wall for the home theater components, place the TV there, being sure to give yourself enough room for wiring the rest of the home theater system. You'll probably want to center the TV along the wall. If you locate it near one corner, you will probably find this placement wastes a lot of space and reduces your installation flexibility. Though not a requirement, all the other components of your home theater system—cable box, VCR, video disc player, and A/V receiver—should be placed close to the TV. This makes connecting everything together easiest, and helps keep cable lengths to a minimum, simplifying wiring.

The placement of the front speakers (left, center, and right) will be largely dictated by the positioning of the TV. That is, the center-front speaker will be placed over or on the TV, and the right- and left-front speakers will be placed from 8 to 12 feet apart flanking either side of the TV. As a good rule of thumb, the front speakers should be placed the same distance apart as the distance from the speakers to the audience. That is, if you are seated eight feet away from the speakers, an ideal separation of the right and left front speakers is about eight feet.

The placement of the subwoofer (if your system has one) and surround speakers requires some additional thought. The subwoofer can be placed almost anywhere in the room, but the most practical location is close to the A/V receiver. This allows the hookup cable to be short. Placing the subwoofer in the middle of the room, or behind the audience, requires a longer cable run and increases the risk of someone tripping over the cable. If the subwoofer is the powered variety, it must be located near an ac outlet.

The surround speakers are commonly placed along a back wall, or two side walls. The cabling for these speakers must be routed to avoid anyone tripping over them (see Chapter 7, "Wiring Your Home Theater System," for ways to hide speaker cabling).

Finally, consider the location of your home theater components and the likelihood of sunlight or other bright light shining on them. TV sets, VCRs, A/V receivers, and video disc players use infrared detectors to receive the signals from their remote controls. The sensitivity of these detectors is diminished when the component is located under bright lights or exposed to sunlight. If you find your remote control doesn't to work very well, try relocating the unresponsive component to a darker area of the room, or block some of the bright light striking the infrared detector. Detectors exposed to the heat of sunlight require a few minutes to "cool down" before they become fully operational.

Typical Hookups

There are an infinite variety of room designs, and thus an infinite number of ways to arrange and connect a home theater system. Still, there are just a few commonly-used, basic hookup schemes that serve as a starting point for your home theater set-up. Be sure to refer to the owner's manuals that came with your equipment. The sample hookup diagrams below provide guidance for inter-connecting your home theater components. Also refer to Chapter 6 for additional information on connecting your system to an outdoor or satellite antenna, and Chapter 7 for details on how to install wiring in your home.

While putting your home theater system together, keep the following points in mind:
- Keep cable lengths as short as possible. It's best to place all the major components of the home theater system in one group. Avoid placing components so far apart that you need cables longer than 6 to 8 feet to connect them.
- Whenever possible, use the video and audio inputs/outputs that provide the best quality. Baseband audio and composite video (phono connectors) provide for adequate quality, but S-Video and component video connections offer greater quality. If your equipment supports these higher-quality inputs and outputs, by all means use them. (For simplicity, we only show standard baseband audio and composite video phono connections in the hookup diagrams that follow.)
- When installing cables, mark each one with masking tape or a cable ID tag (available at Radio Shack). This helps prevent confusion about which cable goes where, and is especially handy if you must replace a piece of equipment at a later date. You'll know at a glance what each cable is for and will be able to reconnect your system with the least amount of fuss.
- If you are installing your home theater system in an entertainment cabinet or center, you may need to pull the cabinet away from the wall so you can access the rear panels of the TV and each component. Remove all video tapes, video discs, books, and other non-essential items before attempting to move the cabinet. Have someone help you if the cabinet is extra heavy.
- While the hookup diagrams in this chapter are useful for giving you an overview of how to connect your home theater system, be sure to refer to the manufacturer's instruction manual before proceeding. The manual will provide instructions specific to the components you are installing and includes important safety and operating tips. After installation is complete, be sure to file these manuals away in a safe, convenient place in case you need to refer to them again.

••• Installing Home Theater •••

BASIC FIVE-SPEAKER SETUP

The minimum home theater system is composed of at least a TV, an A/V receiver with Dolby Surround decoder, and five speakers strategically placed around the room. Three of the speakers are in the front, near the TV, and two are near the back or side of the room. The connection diagram for this basic five-speaker system is shown in *Figure 5-2*. A VCR and cable box also are shown, as these components are usually found in the basic home theater system. Many home theater systems also use a subwoofer to provide extended bass sounds. See "Adding a Subwoofer" later in this chapter for more information about how to connect this type of speaker.

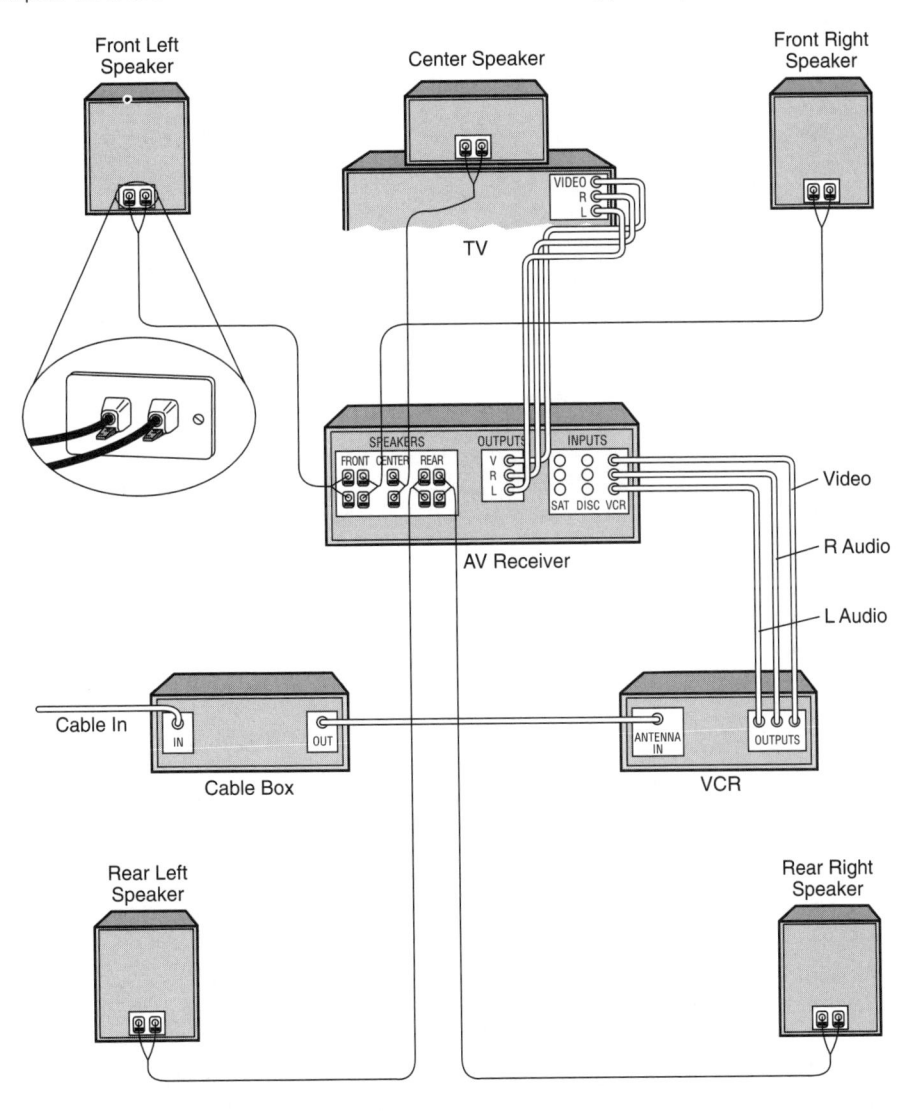

Figure 5-2. The basic home theater setup incorporates a TV, A/V receiver, five speakers, a cable box (or antenna), and a VCR.

Hooking Up the Home Theater System   5-5

SETUP WITH SATELLITE SYSTEM AND OFF-THE-AIR ANTENNA

A satellite dish allows you to receive quality programming directly from space-borne satellites rather than from a local cable television service. The home-based satellite system consists of a dish antenna installed outside connected by coax cable to a receiver located inside, near your home theater system. Satellite systems seldom carry local programming, so it's often necessary to install a separate VHF/UHF antenna outside or in the attic. This system is shown in *Figure 5-3* (refer to *Figure 5-2*, above, for information on hooking up the speakers to the A/V receiver.)

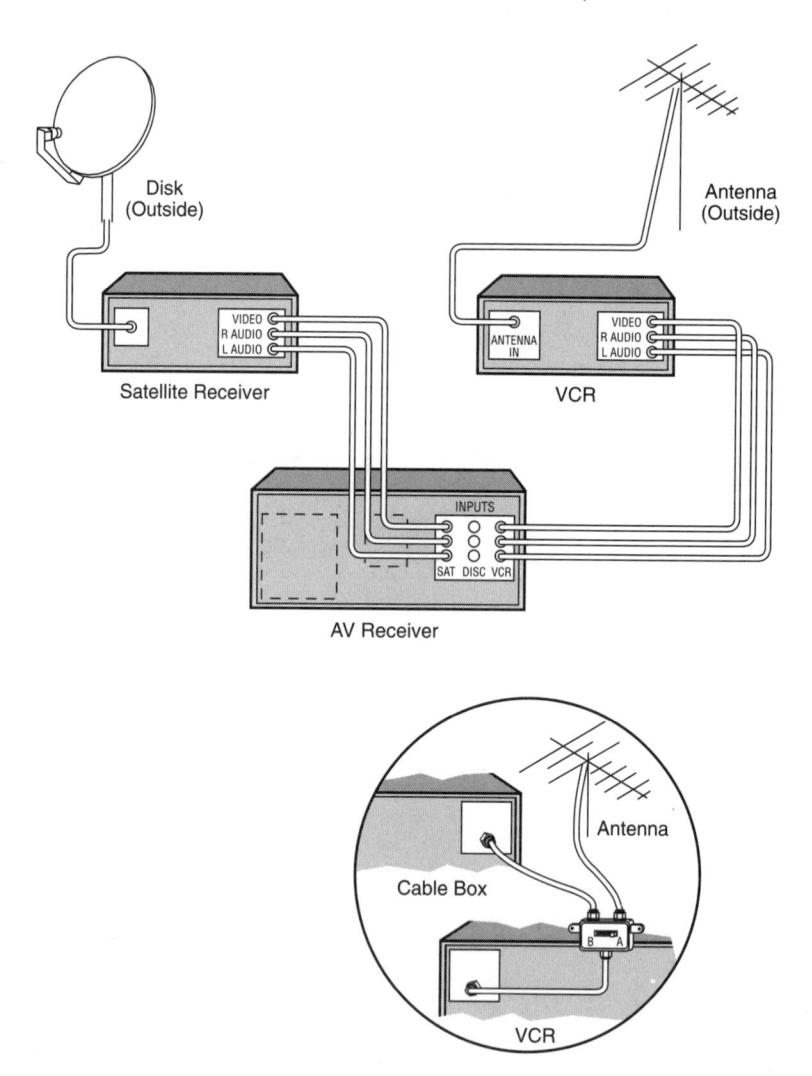

Figure 5-3. The outputs from a satellite receiver can be connected to appropriate inputs of the A/V receiver. The output from a VHF/UHF antenna is an RF signal and must be connected to an antenna input terminal on the VCR or TV. If needed, an A/B switch (see inset) can be used to select antenna or cable inputs.

5-6   Hooking Up the Home Theater System

••• INSTALLING HOME THEATER •••

SETUP WITH VIDEO DISC PLAYER AND VIDEO GAME CONSOLE

The typical VCR offers only limited picture and sound quality due to its engineering. If you want crystal clear pictures and realistic, thundering sound, you need a video disc player. As detailed in Chapter 2, there are two types: 12-inch analog laser discs, and the newer 4.7-inch digital video discs.

Figure 5-4 provides details on hooking up a video disc player and, optionally, a video game console (again, refer to *Figure 5-2* for information on hooking up the speakers to the A/V receiver.) Video game consoles are popular in households with kids. The console is a self-contained computer that plays arcade-quality games stored on disc or cartridge. Many of the latest games come with stereo sound tracks, and a few are even recorded in Dolby Surround! Such games take on an added dimension when played through a home theater system.

Important note! The bright, static images on some video games can cause permanent damage to many models of projection TVs. Be sure to review the instruction manual for your TV to determine if it is the type that can be harmed by a video game.

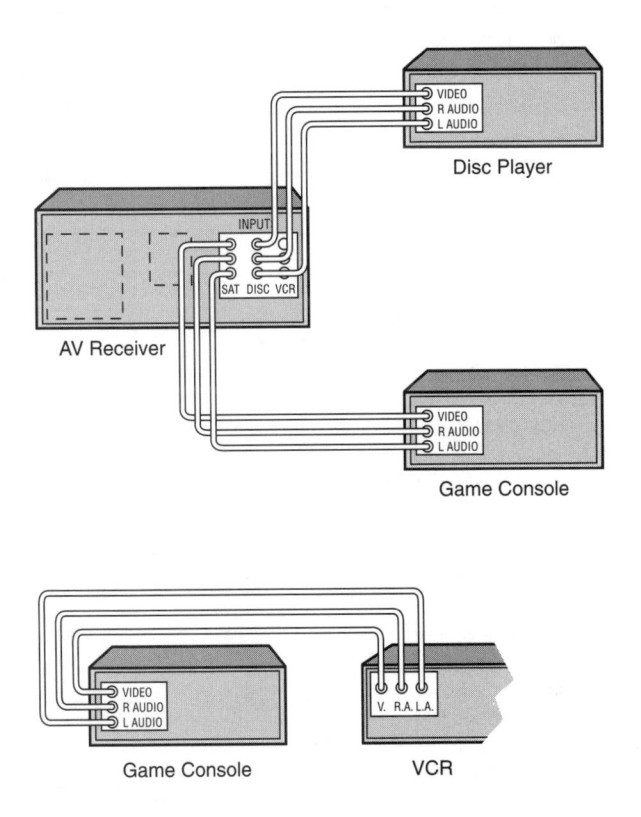

Figure 5-4. Both video disc player and video game console can be connected to unused inputs on the A/V receiver. If there are no available inputs for the game console, you can connect the game console to alternate inputs on the VCR (many have a separate set of "Video 2" connections on the front), or use a separate A/V switcher, discussed later in this chapter.

Hooking Up the Home Theater System   5-7

SETUP WITH REMOTE TV

Many households have more than one TV, but only one home theater system. The home theater system is used as the primary source for television viewing and provides the primary connection for cable, a UHF/VHF antenna, or satellite dish. TVs in other rooms may be used when the full home theater experience is not needed. For example, you may wish to catch an old movie on late-night TV in your bedroom, or watch the news while cooking breakfast in the kitchen.

Figure 5-5 shows two typical ways to add a TV located in another room to a home theater system. In *Figure 5-5a* the remote TV receives the same channel as the one viewed by the home theater system since it uses the VCR as a "rebroadcaster" of whatever is shown on the main home theater screen.

Figure 5-5b demonstrates adding a second cable box or satellite receiver so that the remote TV can tune into a different channel than the one shown on the home theater system. Take note that in order to enjoy this flexibility you need a second cable box or satellite receiver, both of which you will have to rent or buy. When using two satellite receivers you will need a dual-feed LNB on the satellite dish. Read Chapter 6 for more information on satellite systems and their components.

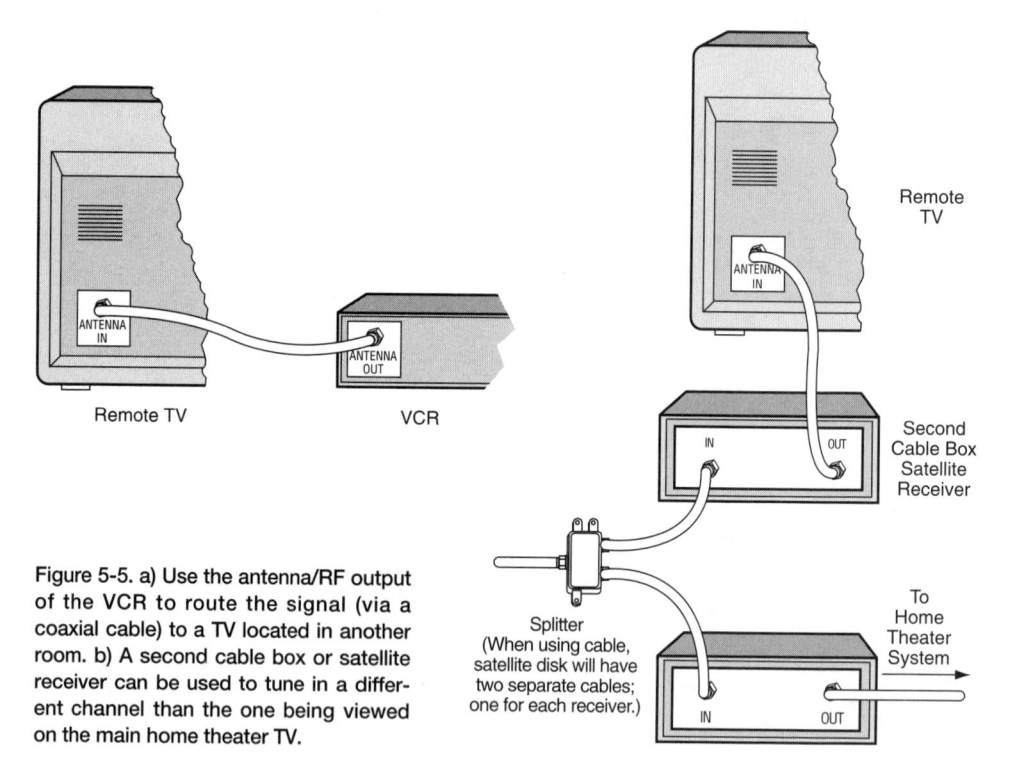

Figure 5-5. a) Use the antenna/RF output of the VCR to route the signal (via a coaxial cable) to a TV located in another room. b) A second cable box or satellite receiver can be used to tune in a different channel than the one being viewed on the main home theater TV.

USING AN A/V RECEIVER

The A/V receiver plays an important role in a home theater system. It not only amplifies the sound for each of the speakers in the system, but also acts as a switcher for selecting one of several program sources. For example, you might decide to watch a movie on tape you've rented, so you switch to "VCR-1" on the receiver. Next, you find a favorite old movie on the satellite system, so you switch to "SAT." Then you select the off-the-air antenna to watch the local, 10 o'clock news. Using the A/V receiver to switch between these sources changes both the audio and the video simultaneously.

Because of the multiple video and stereo audio of inputs, the rear panel of most A/V receivers tends to look complex and intimidating. However, they are actually quite simple: each source (VCR, video disc player, etc.) is provided with a corresponding input to the A/V receiver for video and right- and left-channel audio. Some higher-end A/V receivers provide alternative connector types, such as component or S-Video, in addition to the standard phono jacks. In a typical setup, select the connector type that matches the type of output connection used by the source component. For instance, if your VCR only has phono jacks for baseband audio and composite video, then these are the inputs you use on the A/V receiver.

Some television sets have a built-in surround sound amplifier that can switch between multiple inputs. Terminals on the rear of the TV connect to the external speakers. In such cases, an external A/V receiver is not mandatory, but one is nevertheless recommended since A/V receivers provide greater flexibility and, in general, produce higher quality sound.

ALL ABOUT CONNECTING SPEAKERS

All speakers have an impedance, which is a measurement of their resistance to an alternating current, such as the sound signal from the amplifier in an A/V receiver. The standard impedance for speakers is 8 ohms, but there are specialty speakers available that have 4 and 16 ohm impedances.

Speaker impedance is critical when connecting home theater components—if you use the wrong speaker impedance with your A/V system you run the risk of damaging the speakers, the A/V receiver, or both. This is because the A/V receiver is expecting the speakers to have a certain impedance and drives the speakers with the appropriate amount of power. An impedance mismatch can cause the amplifier to provide excessive power to the speaker, possibly damaging the speaker, the amplifier, or both. Or, the amplifier may deliver too little power causing poor sound quality. Almost all ready-made speakers for home theater systems will be rated for 8 ohms, the same as the standard speaker connection at the rear of your A/V receiver. However, be sure to check that the speakers you use are properly rated for your A/V receiver by comparing the rating label on the rear or underside of the speaker with the A/V receiver.

Speaker impedance also comes into play when connecting speakers together. For example, you may wish to add "remote" speakers in another room by tapping off the main speakers in your home theater system. Doing so changes the total impedance that the speakers present to the A/V receiver, increasing or decreasing it depending on how they are wired together. There are two connection schemes, parallel and serial (shown in *Figure 5-6*), that determine the impedance of the speakers as presented to the amplifier. If you are familiar with basic electronics, you will see that the math used to determine the effective impedance of two or more speakers wired together is the same as that used for connecting two or more resistors together.

SERIES CONNECTION

For speakers connected in series, simply add up the impedance of each speaker. If two speakers are connected in series, and each is rated at 8 ohms, then the effective impedance is 16 ohms. If your A/V receiver is so equipped, use the 16-ohm output on the power amplifier to properly connect these speakers.

PARALLEL CONNECTION

Connecting speakers in parallel reduces their combined impedance. If you are connecting two speakers together in parallel that have the same individual impedance, divide that impedance by two. For example, if both speakers are rated at 8 ohms, the effective impedance is 4 ohms. In addition, if you are connecting together many speakers that have the same impedance, you can quickly calculate the effective impedance of more than two speakers by dividing the impedance by the number of speakers used. For example, if you are using four speakers, and they are all rated at 16 ohms, the effective impedance is 4 ohms.

COMBINING SERIES AND PARALLEL

It is possible to connect multiple speakers both in parallel and in series. You can use this technique, for example, to connect four speakers together, and still present an effective impedance of 8 ohms to the power amplifier. Two of the speakers are wired in series, and two are wired in parallel, as shown in *Figure 5-6*. Assuming all four speakers are rated at 8 ohms, the effective impedance for the *array* of speakers is still 8 ohms.

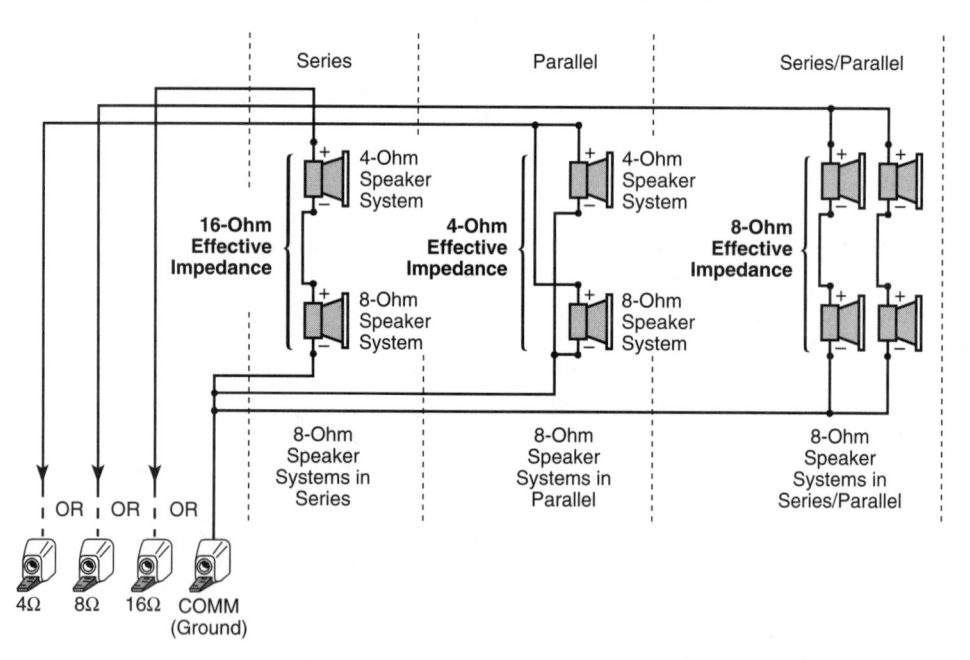

Figure 5-6. There are three possible ways to connect speakers to the output of an A/V receiver. Be sure to calculate the effective impedance of all the speakers no matter what connection scheme you use. For best results, all component speakers should have the same impedance.

••• INSTALLING HOME THEATER •••

INSURING PROPER POLARITY

When connecting multiple speakers, it is absolutely vital that you observe proper polarity at all times. If you mix the polarity, the sound delivered by your home theater system will be substandard. It may sound excessively "boomy," and the output level may be diminished. Refer to the illustrations above for connecting speakers in their proper polarity.

ADDING A SUBWOOFER

It is not uncommon to install a new home theater system but elect to not use a subwoofer. Later, as your home theater tastes mature, you may wish to add a subwoofer. There are two general approaches to adding a subwoofer:

- *Subwoofer added with existing speaker systems.* How the subwoofer is connected depends on whether the subwoofer is amplified or unamplified, and the connections available on the rear of the A/R receiver.
- *Subwoofer added with all-new speakers.* This approach is common with the "subwoofer and satellite" speaker systems, which enjoy a unitized style and smaller size, ideal for apartments or small rooms.

ADDING A SUBWOOFER WITH EXISTING SPEAKERS

Subwoofers are available in two forms: powered or unpowered. The powered versions come with their own built-in amplifier, and are suitable if your A/V receiver provides a "Subwoofer Out" jack. This jack allows you to directly connect the subwoofer to the A/V receiver using ordinary shielded audio cable. Powered subwoofers are desirable if the subwoofer must be located some distance from the A/V receiver (therefore the length of the speaker cable is not a factor), and when there is an outlet available nearby to supply power to the subwoofer. *Figure 5-7* provides a general hookup diagram for a powered subwoofer.

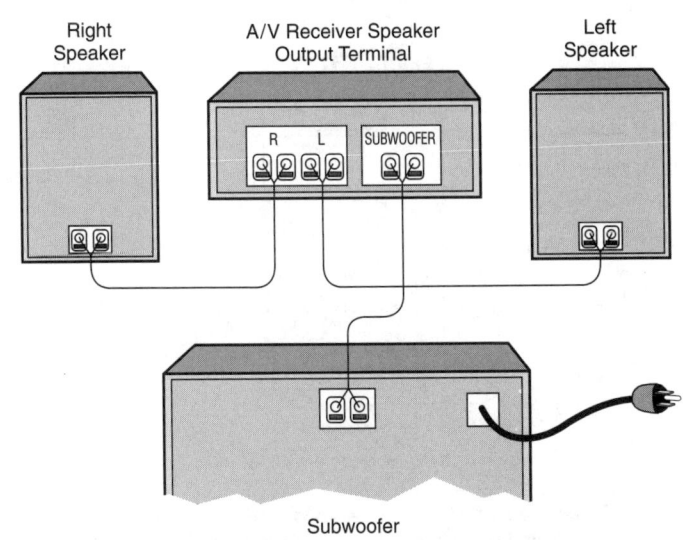

Figure 5-7. Powered subwoofers contain their own amplifier, and so can be connected to the Subwoofer Out (sometimes referred to as Subwoofer Pre-Out) jack on the rear of the A/V receiver.

••• INSTALLING HOME THEATER •••

Unpowered subwoofers are more common, partly because they are less expensive. They don't have a built-in amplifier, and instead derive their power from the amplifier built into the A/V receiver. There are typical connection schemes when using unpowered subwoofers:

- On A/V receivers that have a separate speaker output for a subwoofer, use speaker wire to connect the subwoofer directly to the A/V receiver. See *Figure 5-8a* for details.
- On A/V receivers that lack a separate speaker output for a subwoofer, connect the subwoofer to the terminals used for the right- and left-front speakers, then connect the right- and left-front speakers to the subwoofer. See *Figure 5-8b* for details.

a. Separate Subwoofer Output

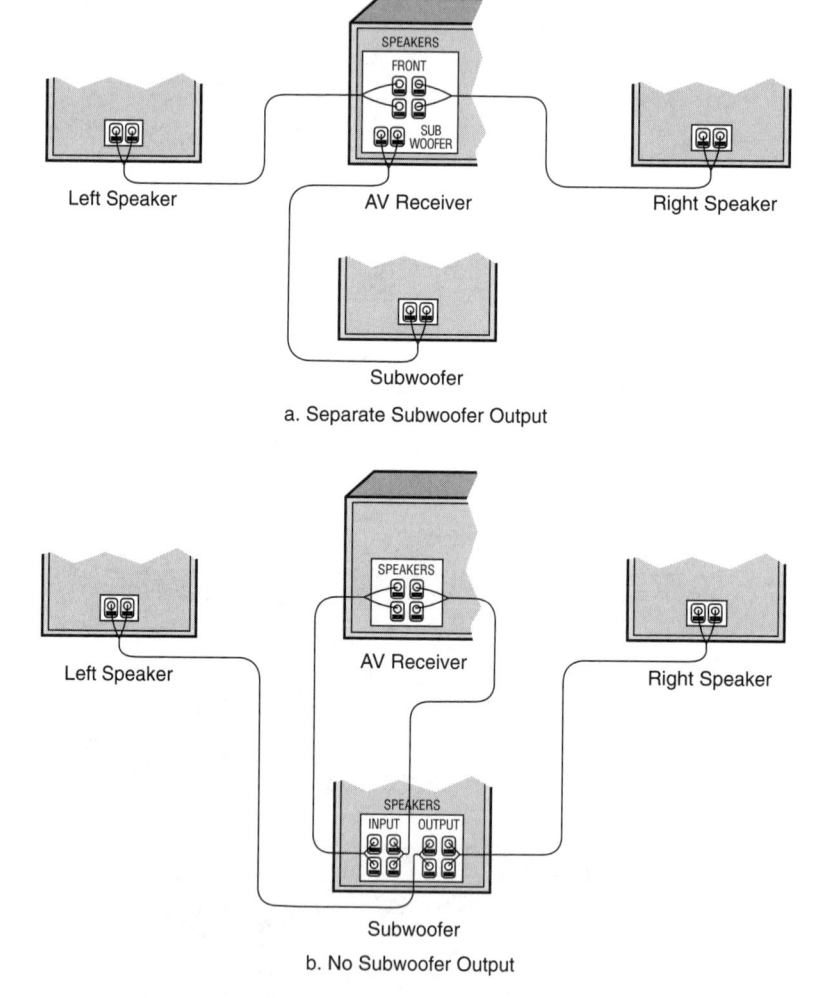

b. No Subwoofer Output

Figure 5-8. a) Some A/V receivers provide their own separate amplification channel for a subwoofer. For these models, attach the speaker wires for the subwoofer directly to the A/V receiver. This is the preferred method for this type of A/V receiver. b) For A/V receivers that lack a separate amplified output for the subwoofer, connect both the right- and left-front speaker terminals to the subwoofer, as shown, then connect to the right- and left-front speakers to the terminals on the subwoofer.

Hooking Up the Home Theater System

••• INSTALLING HOME THEATER •••

ADDING A SUBWOOFER WITH NEW SPEAKERS

When you purchase a new subwoofer you may elect to upgrade all the speakers in your home theater system. If so, consider a home theater speaker set that includes a subwoofer, right- and left-front speakers, and two surround speakers (the set may or may not include a shielded center channel speaker). The benefit of this type of speaker set is that all the speakers are designed to be compatible with one another, both in appearance and in electrical characteristics.

In many cases, a set of this type utilizes a *satellite* speaker design where the subwoofer delivers the bulk of the bass, and individual "satellite" speakers—usually measuring less than 10 inches square—provide the mid- and high-tones. Subwoofer/satellite systems are available in powered and unpowered versions for connection to an A/V receiver. The general approach to hookup is the same as in *Figure 5-8,* above.

ADDING A CENTER SPEAKER

The center-front speaker is often neglected in the planning stage of home theater systems. That's not surprising, because in operation the center speaker should be "audibly invisible" to the audience. That is, viewers should not even be aware of its existence. The sound should appear to magically come from the TV screen itself.

Many higher-end television models incorporate their own built-in, quality speakers that can be used as the center speaker in your home theater system. There are two ways to connect the A/V receiver to the speaker in the TV: via the amplified speaker terminals, or the unamplified center speaker output. The amplified speaker terminal is used if the TV set provides terminals for direct access to the speakers (not all TVs do). This method is desirable because you can set the overall sound level from the A/V receiver, ignoring the volume control on the TV. The other method is to connect the speaker in the TV to the unamplified output jack on the A/V receiver (usually marked Center Out or Center Pre-Out). Since you are using the TV's own amplifier when using this approach, you will need to adjust the volume on the TV to change the output level of the center speaker.

The speakers that are built into most TVs, even on high-end models, are not always as good as separate center-channel speakers, so you may wish to add a separate center speaker to improve sound quality. Like subwoofers, you can buy powered and unpowered center channel speakers. Unpowered models are more common, and connect to the center-channel speaker output terminals on the back of the A/V receiver. If you use a powered center-channel speaker, connect it to the Center Out (or Center Pre-Out) jacks on the rear of the A/V receiver.

RUNNING SIGNAL CABLES AND POWER CORDS SEPARATELY

Even though your home theater system uses shielded cables, stray electrical noise can enter the cable and be heard through the speakers. One of the most annoying electrical noises is hum, which usually is caused by the proximity of power cords to signal cables. The hum is created by the 60 Hz (cycle) alternating current flowing through the power cord that is inductively transferred from the power cord to the signal cable.

While shielded cables (and connectors) certainly do their share of keeping stray electrical noise from entering your system, you can prevent most induced hum simply by keeping power cords physically separated from the signal cables. Whenever possible, avoid placing power cords and signal cables side by side. As a rule of thumb, keep them apart six inches or more, especially if the power cord and signal cable are routed parallel to one another. And never let loops of cords and cables near one another. The loops act as transformers, and a very noticeable amount of hum can result.

ATTACHING AN F-CONNECTOR

As mentioned previously, you can either buy manufactured video cables made to the approximate length you need with the proper F-connectors already attached, or you can make your own. Making your own cables can save money and they can be cut to the exact length for the job. *Figure 5-9* is a step-by-step guide showing how to attach an F-connector onto coaxial cable. Please note that F-connectors are designed for a specific cable type—for example, connectors designed for RG-59 cable won't fit on RG-6 cable.

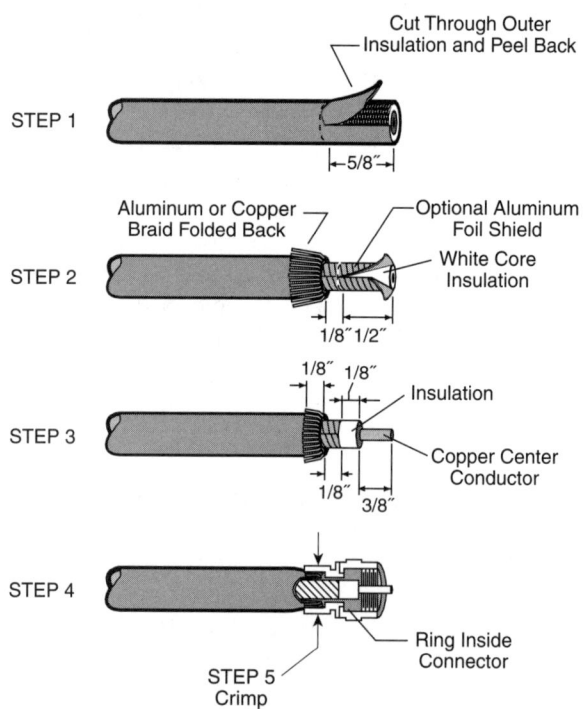

Figure 5-9. You can make your own F-connectors by carefully trimming the insulation and dielectric of a coaxial cable, as shown here. F-connectors are available in screw-on or crimp-on varieties.

Steps:
1. Strip the outer jacket insulation to expose about 7/16-inch of the braided wire. At this point, don't cut the braid.
2. Fan out the braid, fold it back over the cable jacket, and trim to about 1/8-inch in length.
3. Some cables have an aluminum sheath under the braid. Strip off the excess sheath and foam insulation to expose 1/4-inch of the center conductor—do not

cut too deeply into the dielectric or you will nick the center conductor. Make sure the center conductor is clean and bright-looking.
4. Slide the F-connector over the foam insulation and *under* the braid. Push the fitting onto the cable until it seats firmly and no braid is showing. If the fitting is the crimp type, crimp it with the proper tool.

TOOLS LIST

Although specialty tools aren't absolutely necessary for attaching F-connectors to coaxial cable, using the right tool makes the work go faster and gives a more professional-looking result. You can cut the foam insulation around the center conductor with a knife or use a conventional wire stripper. The stripper helps prevent nicking the conductor. The best method is to use a coax cable stripper tool, which works with all sizes and varieties of coax, and automatically cuts off the proper amount of outer insulation and foam.

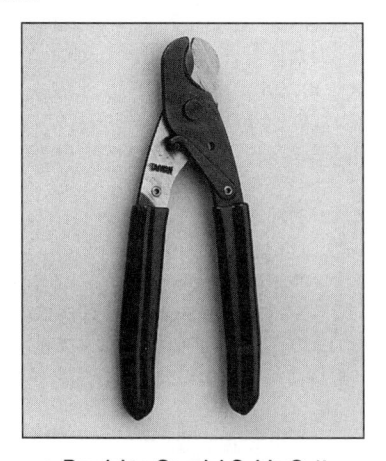

a. Precision Coaxial Cable Cutter

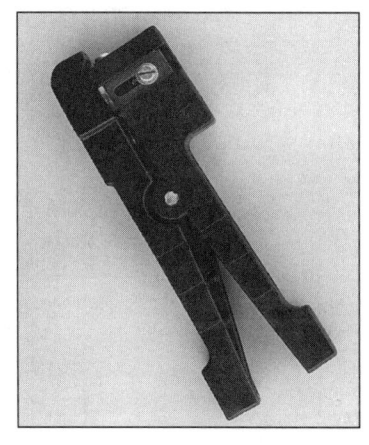

b. Coaxial Cable Stripper

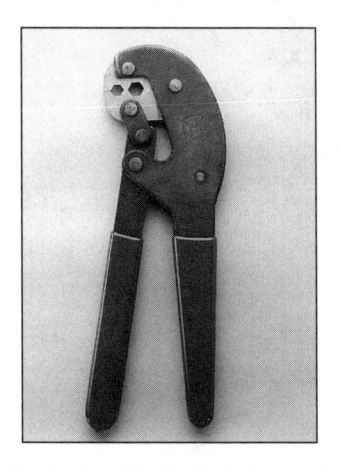

c. Professional-Style Hex Crimping Tool

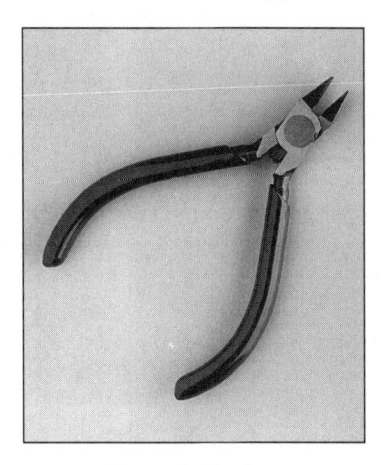

d. Diagonal Wire Cutters

Figure 5-10. Special tools for coaxial cable installation.

Hooking Up the Home Theater System   5-15

••• INSTALLING HOME THEATER •••

Because F-connectors are large, they require a special tool for proper crimping (regular crimpers don't work well because they are made for smaller wire and connectors). The tool is available in a variety of styles—your choice from economy to heavy-duty professional models (prices range from about $3 to $15).

If the cables will be used indoors, you can use crimpless F-connectors. These screw securely onto the end of the cable. Crimpless connectors are about twice as expensive as their crimped cousins, so you may not want to use them extensively.

UNDERSTANDING SIGNAL IMPEDANCE AND RESISTANCE

All signal cables and wires are subject to two basic electrical behaviors: impedance and resistance.

IMPEDANCE

Impedance is the electrical opposition encountered by an alternating current and is expressed in ohms. You will commonly see impedance specified for speakers and for wire. Speakers present a "load" to a circuit; the load is the work the circuit must perform to produce some output. The typical impedance for speakers is 8 ohms, which means the speaker presents a load of 8 ohms to the power amplifier of your stereo or A/V system. If you use the wrong impedance, you risk damaging the amplifier, the speaker, or both.

Impedance also is a factor in cables and wires because using the wrong impedance can degrade signal quality. In modern video and home theater systems, RF signals are transmitted through coaxial cable. Video systems for the home are engineered for 75 ohm impedance. The coaxial cable used to connect these components must be rated at 75 ohms. Otherwise, the signal can be degraded, exhibiting "ghosting" and a loss of overall signal strength. See "Which Coax?" below, for more information.

RESISTANCE

Resistance is the electrical opposition encountered by a direct current. Like impedance, resistance also is measured in ohms. While carrying video and audio signals to and from your VCR, the resistance of a cable roughly determines how much of that signal reaches its destination. Resistance plays an important role in speaker wire. The more resistance encountered in speaker wire, the lower the sound level from the speakers. Resistance in wire is directly related to the diameter of the wire. Small diameter wire, such as 18 or 22 gauge, is not suitable for use in home theater systems as it cannot conduct adequate current to power the speakers. Sixteen gauge is the minimum size you should use for speaker wire; heavier 12- or 14-gauge wire, such as Radio Shack Megacable®, is recommended for high wattage applications, such as front channel speakers or subwoofers.

Higher wattages and longer runs of wire demand larger wire. *Table 5-1* summarizes recommended wire gauges, assuming 8 ohm speakers. A maximum 100 foot wire length is also assumed.
- When using 16 ohm speakers, the recommended minimum wire size is one gauge smaller—e.g. instead of 14 gauge wire, it is acceptable to use 16 gauge wire.
- When using 4 ohm speakers, the recommended minimum wire size is one gauge larger—i.e. instead of 14 gauge wire, you should use 12 gauge wire.

Speaker wire larger than 12 gauge is available, but is difficult to find and much more expensive. Rather than use a larger wire, it is generally more economical and practical to simply reduce the wire length. This can only be done by placing the power amplifier closer to the speakers.

Maximum Wattage	Recommended Minimum Gauge
25	16
50	14
100	12

Table 5-1. Recommended Speaker Wire Gauge (maximum 100 foot length).

Which Coax?

Coaxial cable is available in a number of different impedances. The proper impedance for video is 75 ohms. Both the RG-6 and RG-59 coax cable, available at Radio Shack and most any TV, video, or electronics store, is rated at 75 ohms. RG-6 is heavier and uses a larger center conductor than RG-59. RG-6 is considered the better, all-around choice when using long cable lengths because there is less signal loss per 100 feet of RG-6 than with RG-59 (see *Table 5-2*).

However, RG-6 has a higher capacitance per foot that RG-59 (see the table below). So if you're using short lengths of cable—less than four or five feet—the better choice is RG-59. The best news: RG-59 is less expensive and generally easier to work with. Note that most F-connectors are made for RG-59. If you are using RG-6, be sure to get the proper F-connectors with the larger crimp barrel.

Remember, not all coax is designed for use with video, so the impedance of other coax types may not be 75 ohms. Two commonly available coax cables designed for radio use (CB, ham) are RG-8 and RG-58. These have an impedance of only 52 ohms. Using either of these coax cables with your video system will impair the quality of the sound and picture.

Signal loss in a cable is expressed in decibels (dB) at different TV broadcasting frequencies. Every three dB is a doubling in signal loss. Note that signal loss increases at higher frequencies.

Type	Outside Diameter	Signal Loss / 100 Feet	Center Conductor
RG-59	0.242"	1.8 dB @ 50 MHz	22 gauge
	0.242"	2.8 dB @ 100 MHz	22 gauge
	0.242"	7.5 dB @ 500 MHz	22 gauge
RG-6	0.266"	1.8 dB @ 50 MHz	18 gauge
	0.266"	2.8 dB @ 100 MHz	18 gauge
	0.266"	7.5 dB @ 500 MHz	18 gauge

Table 5-2. Typical Specifications of Common 75-ohm Coax Cable

••• INSTALLING HOME THEATER •••

CALIBRATING THE SPEAKER OUTPUT

The typical home theater system uses five speakers placed strategically around the room. The sound level of these speakers must be carefully matched, or the surround sound effect will be diminished. A quick way of "calibrating" the speakers is by ear.

CALIBRATING "BY EAR"

Most A/V receivers equipped with Dolby Pro-Logic or Dolby Digital circuits incorporate a Test button for use in calibrating the home theater system. Depressing the Test button produces a sound tone through each speaker in sequence.

- For Dolby Pro-Logic systems (monophonic surround speakers), the sound will alternately come from the left-front, center, right-front, and both surround speakers.
- For Dolby Digital systems, the sound will alternately come from the left-front, center, right-front, right- and left- surround speakers. A test sound may or may not be heard from the subwoofer.

To use the Test button:

- Adjust the volume control on the A/V receiver to a normal listening level.
- Depress the Test button, then note the loudness of each speaker. The sound level from each speaker should be approximately equal.
- Adjust the controls on the A/V receiver to increase or decrease the level from each speaker, as needed.
- Once the calibration is complete, depress the Test button again to turn off the test sound.

If your A/V system lacks a Test button, you can conduct a calibration using a test tape or disc. Test tapes and discs often can be found at stereo stores or by mail order. Note that the better test tapes and discs can be expensive since they are made to exacting standards. When using the test tape, switch to true Dolby surround on the A/V system. Do not test and calibrate the system using another surround setting. For example, many A/V systems can synthesize multi-channel sound. Test tapes cannot be used adequately with synthesized sound sources.

If you do not have a special test tape, use a commercially prerecorded video tape or disc of a program that is known to rely heavily on Dolby surround sound effects. Most any recent action-adventure film contains sequences that can be used to adequately test your home theater system.

During calibration, make sure that the level for the center-front and surround speakers is properly adjusted. The surround speakers should not be set too loud or they can become distracting. The level of the center speaker should be high enough so that all on-screen dialog appears to come from the screen and not from the right- and left-front speakers. The front speaker should seldom be "dead" and not delivering any sound while the other speakers are active. This can create an unnatural sound "hole" where the sound does not appear to come from the TV screen.

CALIBRATING BY SOUND LEVEL METER

While the "ear test" is certainly satisfactory for the typical home theater system installation, a more precise method of testing the sound levels in a room is to use a sound level meter. Sound level meters are inexpensive testing devices designed for one purpose only—to indicate the relative volume of sound at a given location.

••• INSTALLING HOME THEATER •••

The design of the sound level meter is fairly simple: the meter consists of a sensitive microphone connected to a precision amplifier. The output of the amplifier is attached to a meter device with either an analog needle or digital readout. The meter face is marked in dB SPL (which stands for sound pressure level). When sound enters the microphone, it is amplified and the strength of the signal is read on the meter.

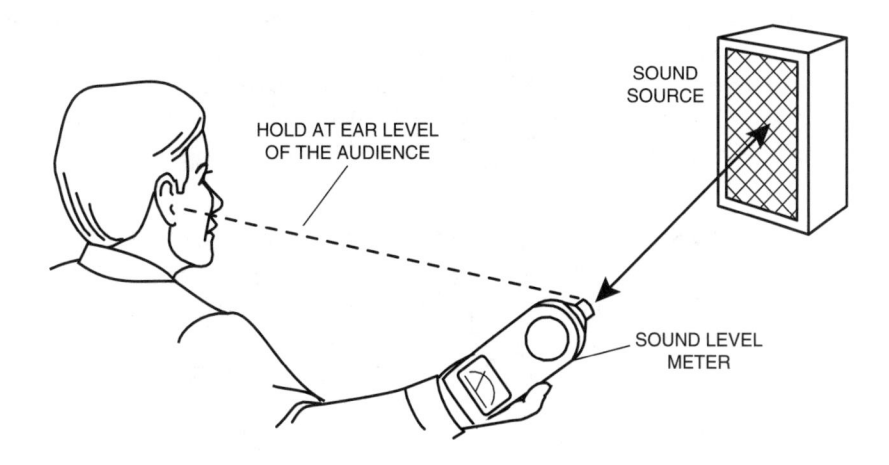

Figure 5-11. Use a sound level meter to accurately test the level from each speaker in your home theater setup. When used with a test signal generated by the A/V system, the level from each speaker should be about equal.

Sound level meters also incorporate various controls that you can use to tailor the measurement.
- A dial lets you specify the range of the sound level. You adjust the range to the highest peak you anticipate (example: 70 dB if you expect no sound over 70 dB). Setting the range to just about the maximum output level improves the accuracy of the meter.
- A weighting switch lets you adjust between A and C weighting. A weighting closely approximates the hearing characteristics of the human ear, where higher frequencies are heard more readily than lower frequencies. C weighting is a more linear measurement, where all the frequencies are considered. In most cases, switching the meter to C weighting will result in a slightly higher reading, because the very low frequencies are not ignored. The type of program influences the weighting that should be used. Rock music is often best measured using the C weighting; other programs can be measured using A weighting.
- A response time switch lets you adjust the rate at which the meter takes its measurements. A fast response time indicates the instantaneous sound level, where a slow response time indicates an average sound level, taken over a period of perhaps as much as a second. Use a fast response time if the program material is very "active" and has lots of peaks, like loud drum beats. Slow response is perfectly suited for softer music or voice.
- A hold button lets you "freeze" the action of the meter at any given time. You might use the hold button, for example, when holding the meter over your head,

where you can't easily see the meter reading. Press the hold button, then read the meter.
- A peak switch shows you the loudest measurement taken by the meter. This is helpful if you want to determine SPL of the loudest portion of the program.

Position yourself in the audience seating area of the room. Place the meter approximately at ear level, but away from your body. The meter should point directly at the source. Press the Test button on your A/V system and note the reading on the meter as the test tone or sound is heard from each speaker. The meter should read about the same level from each speaker. Note that while you should strive for equal readings from all the speakers, an absolute match is nearly impossible to achieve and is seldom necessary. A variance of one decibel will not be noticeable, but a variance of three decibels will.

And Now: Wiring for Satellite Dishes and Antennas

This chapter provided details on connecting the components of a home theater system. You learned how to attach multiple programming sources to your home theater system, and how to attach the speakers to the A/V receiver. But, your home theater system may not yet be complete. Your system may be augmented with a satellite dish or outdoor antenna, both of which require special installation. The next chapter discusses the how and why of choosing and installing a satellite dish and/or outdoor antenna for use with your home theater system.

Satellite Dish & "Off-the-Air" Antenna Hookups

Despite the popularity of cable television, some 35 million homes in the United States continue to get their TV broadcasts using an "off-the-air" antenna, typically installed on their roof or at the side of their house. And millions more now receive their programming via direct-to-home satellite, aiming a parabolic dish antenna skyward to intercept the shower of channels beamed to earth from 23,000 miles into space.

If you use an outdoor antenna, even for part-time viewing, you'll want to make sure yours has been properly selected for the channels you receive. If you use an antenna and it's old or poorly installed, you could be cheating yourself out of some quality TV viewing. With the right antenna, properly installed and maintained, your TV can pull in distant stations, and even thaw snowy pictures. In some cases, an "off-the-air" TV antenna may provide clearer pictures than those offered by cable, or additional local programming that is not carried on a satellite system.

This chapter details the selection and installation of both satellite dish antennas and outdoor, off-the-air antennas. Installation details for satellite antennas is generic, since different satellite programming services entail different types of satellite antennas and, therefore, different antenna installation procedures.

Receiving Satellite Television

The fastest-selling consumer electronics product in history has been the mini-dish direct broadcast satellite (DBS) TV reception system. These systems provide viewer access to more than 200 TV channels without connection to a cable TV system.

Home television reception via satellite has been available since the late 1970s, when program providers such as HBO and Showtime® turned to outer space to distribute their services. Back then, the basic home satellite system consisted of a large (usually 12-foot) metal or fiberglass dish, and a home-built or commercially-made satellite signal receiver. This setup usually cost $5,000 or more, but many hobbyists considered it a worthwhile investment, since the programming was free.

Soon, HBO, Showtime, The Disney Channel®, and other program providers began to "scramble" their signals in order to prevent them from being received at no cost by backyard satellite dish owners. With the aid of several laws that guarantee access to satellite programming, manufacturers developed home-based versions of satellite signal descramblers. Since then, home satellite—also called direct-to-home or DTH

satellite—has become a substantial business. Today, there are two general forms of DTH satellite services:
- The newer DBS system uses a small, 18-inch satellite dish antenna to receive all channels from a single, high-powered satellite. Usually, programming "packages" are purchased through one or two providers, similar to cable TV. This technology uses digital transmission and signal compression techniques that allow more than 200 channels to be broadcast from a single satellite.
- "Big dish" systems, often referred to as Home Satellite Systems (HSS), use larger dish antennas. In this technology, each satellite is able to broadcast only about 24 channels, so the antenna must be able to receive signals from more than one satellite. Generally, this requires the ability to move the dish and aim it to receive signals from each satellite. Programming may be purchased through any of several independent providers. Today's HSS antenna is a direct descendant of the home satellite gear from the 1970s. Modern HSS antennas are generally smaller (7 1/2 feet is the norm, rather than 12 feet), thanks to improvement in satellite and electronics technology. The satellite receivers also have become much more sophisticated, but easier to use.

There are pros and cons to both forms of satellite reception, as summarized in *Table 6-1*. Only you can decide which type of satellite system you wish to get.

Factor	DBS	HSS
Size	Smaller overall size allows installation almost anywhere, including the window sill of an apartment building.	Larger dish size is most suitable for single-family homes with large yards in areas where antenna restrictions do not apply.
Installation	Unitized parabolic dish antenna and LNBF is easier to install. Installation kits help speed the process. Antenna position must allow for clear line-of-sight to only one satellite.	Installation is more difficult because of movable nature of dish antenna. Must be professionally installed in most cases. Antenna position must allow for clear line-of-sight to a number of satellites across the southern sky.
Program availability	Good, but programming selection is determined by program provider(s).	Excellent. As new programs are available on satellite most are available immediately to big-dish owners.
Initial cost	Low, under $300 for most systems.	High, $1,500 and up for a typical system.
On-going cost	Programming costs are similar to cable television service.	Lower overall costs due to competition between program providers—often 25 to 50% less than DBS.
Upgrades	Receiver and dish unit (including LNBF) can be separately replaced and/or upgraded.	All components, including receiver, LNB, feedhorn, and dish antenna can be separately replaced and/or upgraded.

Table 6-1. Comparison of DBS and HSS satellite TV antenna systems.

INSTALLING A DBS SYSTEM

DBS systems are the most popular form of satellite TV reception, thanks to their small size and relatively low overall equipment costs. There are several DBS systems to choose from. One alternative is to purchase your own equipment, such as the Ra-

••• Installing Home Theater •••

dio Shack Optimus® or RCA system. Alternatives include systems that can provide different signals to two or more TV sets. You can even buy systems for mobile homes, as well as portable units. Having purchased the equipment, you then subscribe to programming providers.

An alternative to purchasing the equipment is offered by Primestar®, which is a service that installs the dish antenna and provides the programming. There is a one-time installation fee and a monthly fee for programming packages. Except for slight variations in antenna dish size, most systems are comparable; however, they are not interchangeable. If you purchase a system, you cannot receive Primestar services, and vice versa.

DBS systems come with, or have available as an option, kits for installing the antenna on the roof or side of your house. The kit is highly recommended, as it comes with all of the hardware—and step-by-step instructions—required for typical antenna installation. If you prefer not to do the installation yourself, you can have someone else do the work for you. The installation may be an extra charge, or it may be part of the purchase price of the system. Be sure to ask.

DBS systems include these major parts, as shown in *Figure 6-1*.
1. *Satellite dish antenna.* This "antenna" is actually a parabolic reflector (18 inches for DSS; 18-36 inches for Primestar and other systems) that collects the Ku-band (10.95-12.7 GHz) signals beamed from the satellite.
2. *Low-noise block downconverter feedhorn, or LNBF.* This is the actual receiving element of the antenna. The LNBF also converts the very high frequency microwave signals used in satellite transmission to a form more readily transmitted through the cable into your home.
3. *Receiver.* The set-top receiver has all of the electronics built into it to decode the scrambled satellite signal. The receiver tunes to the one channel you wish to view out of the hundreds available through the satellite broadcast service. The receiver is connected to your A/V receiver, VCR, or TV, as depicted in Chapter 5.

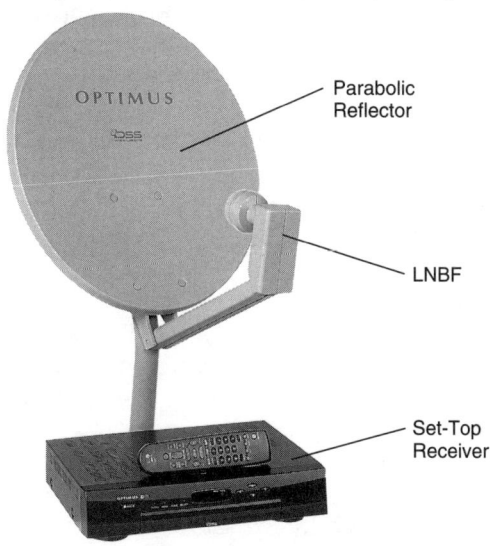

Figure 6-1. The basic DBS system is composed of a dish antenna, an LNBF, and a set-top receiver with remote control.

••• INSTALLING HOME THEATER •••

In general, these are the steps to follow when installing a DBS system:
1. Install dish antenna on outside of the house. Three methods are commonly used: eve mount, chimney mount, and side mount. In all cases, the dish antenna can only be placed where there is proper exposure to southern skies. Since DBS systems receive all of their signals from one satellite and don't use a motorized antenna, *only that portion in direct line-of-sight between the DBS dish antenna and the satellite must be obstruction-free.* Trees, buildings, and other obstructions will block the relatively weak signals from the satellite and you will not be able to receive programming.
2. Route the antenna cable from the dish to your home theater system. See "Routing the Lead-in Cable," later in this chapter, for some tips on running the cable into your house. Also see Chapter 7, "Wiring Your Home Theater System," for more information.
3. Inside the house, attach the antenna cable to the satellite receiver.
4. Connect the receiver to your home theater system. The suggested method is to connect the audio and video outputs of the satellite receiver to appropriate audio/video inputs of your A/V receiver.
5. Turn the receiver on and select the Test or Setup mode. Follow the instructions provided by the manufacturer to aim the satellite dish for maximum reception signal strength. Having a "helper" move the dish up and down, and left to right, will greatly aid in aligning your satellite dish.
6. Once the dish has been installed and aligned, you may purchase or complete the purchase of the programming you wish to receive.

INSTALLING AN HSS SATELLITE SYSTEM

HSS (Home Satellite System) is a general term for DTH satellite systems that receive C-and Ku-band satellite signals. Most, but certainly not all, of these systems are designed to receive programming from more than one satellite. The satellite dish antenna usually has a motor (called an *actuator*) that turns the dish along an axis to point it at various satellites. Because HSS systems are designed to receive signals from many satellites, they are more difficult to install and align. For this reason, you may wish to have an experienced installer set up your big-dish system.

An HSS system is comprised of the following components.
1. *Satellite dish antenna.* This antenna, actually a parabolic reflector, measures from 5 to 10 feet in diameter (the average size is 7 1/2-feet). Larger sizes are often needed for those living in Alaska, Hawaii, the far northwest and northeast United States, and in Florida.
2. *Low-noise block downconverter, or LNB.* This is the actual receiving element of the antenna. It also converts the very-high microwave frequencies used in satellite transmission to a form more readily transmitted through the coaxial cable into your home.
3. *Feedhorn.* This device collects the signals reflected from the parabolic dish and focuses them into the LNB. It also has a polarizing element to properly receive the channels transmitted by the satellite. In some systems, the LNB and feedhorn are housed in the same unit.
4. *Receiver.* The receiver tunes one channel out of the 24 channels relayed by the

••• INSTALLING HOME THEATER •••

satellite. Each satellite can beam multiple channels, providing for hundreds of channels overall. The receiver is then connected to your A/V receiver, VCR, or TV, as depicted in Chapter 5.

Most programming for HSS systems is scrambled so that the program provider can earn money from home satellite dish subscribers. You pay a monthly or annual fee to receive a channel, or set of channels. Unlike DBS systems, the electronics for descrambling channels is not an integral part of the receiver of an HSS system. Rather, descramblers are add-on modules that are inserted into the receiver. There are two types of descrambling modules available for HSS systems:

- The VideoCipher® module descrambles analog satellite signals. Most satellites support up to 24 analog channels at a time. The channels are transmitted through separate *transponders* on the satellite. All modern satellite receivers are built to accept the VideoCipher (also known as VCII or VCRS) module.
- The DigiCipher® module descrambles digital satellite signals. Multiple digital signals can be compressed onto one satellite transponder, allowing a single satellite to carry several dozen programs. Only the latest satellite receivers are built to accept the DigiCipher module. These typically carry the branding "4DTV." Newer units are able to descramble both analog and digital signals.

Note: Some satellite programming, particularly religious and "shop-at-home," is not scrambled. It can be received on your big-dish system without the use of either the VideoCipher or DigiCipher module.

In general, these are the steps to installing an HSS system:
1. Assemble the dish antenna, mount, and other components, as specified in the manual that came with your system. Most satellite dishes are shipped in pieces or "petals" and you build the dish by connecting the petals.
2. Install the dish antenna outside of the house. Installation can be rooftop or ground-level. Rooftop installations are best if the antenna is under 8-feet, and only when it is a wire mesh dish design, which provides less wind resistance and snow load. Ground-level installations require that a pole be secured into a cement footing. *In both cases, the dish antenna can only be placed where the line-of-sight view to the entire arc of satellites in the southern sky is unobstructed.*
3. Rough-align the dish so that the dish mount is pointed due north, (the dish itself will be pointed to the south) and that the "elevation" of the dish (its angle relative to the ground) is appropriate for your latitude.
4. Route the antenna cable from the dish to your home theater system. See "Routing the Lead-in Cable," later in this chapter, for some tips in running the cable into your house. Also see Chapter 7, "Wiring Your Home Theater System," for more information.
5. Inside the house, attach the antenna cable to the satellite receiver.
6. Connect the receiver to your home theater system. The suggested method is to connect the audio and video outputs of the satellite receiver to the appropriate audio/video inputs of your A/V receiver.
7. Turn the receiver on and select the Test or Setup mode (not all receivers offer these modes). Follow the instructions provided by the manufacturer to aim the satellite dish for maximum reception. Have someone help you adjust the mounting of the dish so that it properly tracks all satellites from east to west. This is a

difficult task for most people, and may require several hours to get it right.
8. Once the dish has been installed and aligned, you may purchase or complete the purchase of the programming you wish to receive.

How Off-the-Air Antennas Work

The word *antenna* is derived from Latin words that mean "sail yard." Instead of catching the wind at sea, off-the-air television antennas are designed to catch the minute radio wave signals broadcast from a TV tower.

Let's start with how a TV signal gets from the station's transmitter to your TV set. All TV signals (radio and other electromagnetic signals as well) begin as an electrical current in the station's transmitting antenna. This current creates an electromagnetic field around the antenna and that field radiates out from the transmitting antenna like waves on a pond when a rock is thrown into the water. These "radio waves" travel at the speed of light (about 186,000 miles per second) and, at lower frequencies, can travel vast distances. TV signals are very high frequency (VHF) and ultra-high frequency (UHF) and travel in line-of-sight.

Just as radio waves are created by electric currents flowing into the TV station's transmitting antenna, the opposite occurs in a receiving antenna. As the transmitted radio waves cut across a receiving antenna, they induce a voltage in the antenna which creates an electric current that is the mirror image of the signal coming from the transmitting antenna. These signals are then processed by your TV set and one channel is selected for your viewing. To learn more about how TV and other types of antennas work, read *Antennas* by A. Evans and K. Britain, published by Master Publishing, Inc., and available from Radio Shack (RS # 62-1083).

TYPES OF OFF-THE-AIR ANTENNAS

Outdoor and indoor TV antennas for local broadcasts come in many forms, and each one has its own unique application. Some are designed to intercept VHF broadcasts (channels 2 through 13); while others are tuned for catching the UHF signals (channels 14 through 69). Still others are constructed for all-purpose reception, capturing VHF, UHF, and FM radio broadcasts. Though there are many types of antennas used to receive television broadcast signals, most can be divided into two main groups: dipole and yagi.

The Basic Dipole Antenna

The basic TV antenna is the simple *dipole*. It is composed of two rods separated by a center insulator. A pair of wires connect your TV to the rods. The length and diameter of the rods depends on the frequency of the TV station you want to intercept. An improvement on the simple dipole is the folded dipole. This antenna uses two dipoles in parallel (side-by-side) with their ends connected. The signals are picked up by each dipole and added together. An example of a dipole antenna is shown in *Figure 6-2*.

••• INSTALLING HOME THEATER •••

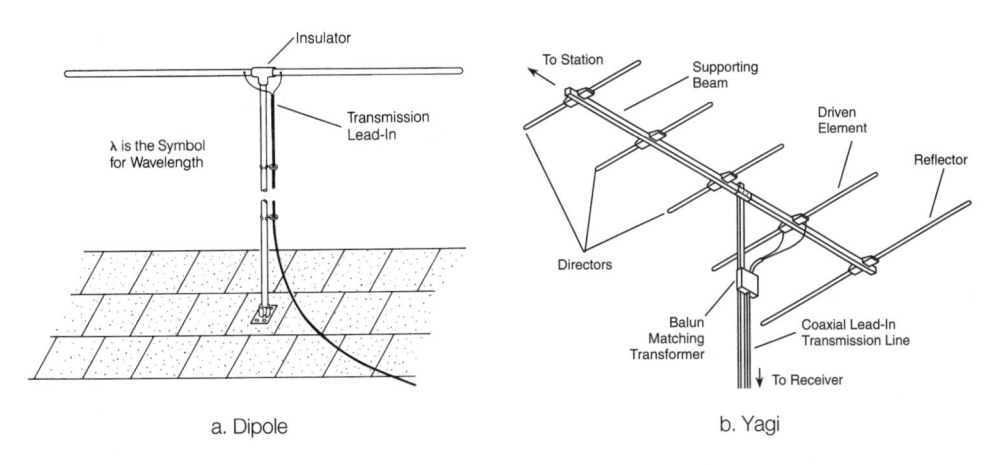

a. Dipole

b. Yagi

Figure 6-2. a.) The basic dipole antenna uses two rods separated by a center insulator. The folded dipole is essentially two dipoles in parallel with their ends connected. b.) The Yagi antenna is a dipole antenna with directors (the short elements) and a reflector (the longest element) to improve signal reception.

The Yagi Antenna

The most common outdoor or attic TV antenna uses several elements similar to dipoles (simple, folded, or both) arranged in single-file on a metal boom. Only one of the dipoles, called the driven element, is electrically connected to the TV set. The others serve as *directors* and *reflectors*. The directors help "zero in" on a signal, and the reflectors help collect extra signal to boost the reception by reflecting the signal back to the driven element. This type of antenna is called the *yagi*. It was invented by a Japanese professor, Mr. Uda, but is named for the person who filed the English translation describing the antenna, Hidetsugu Yagi, a Japanese physicist.

The yagi has two chief benefits: gain and directionality. Gain is like amplification, even though an antenna doesn't use any electronic circuitry. It means that the antenna collects more signal from a particular direction that a simple point-source antenna. The more gain, the stronger the picture. Directionality is the ability of the antenna to concentrate or focus on receiving signals within a narrow beam coming from local and, especially, distant locations.

SPECIAL CONSIDERATIONS FOR UHF CHANNEL RECEPTION

The yagi is the most popular antenna used to receive UHF signals. To increase the signal-capturing ability of the antenna, the yagi design usually incorporates a "corner" reflector to pick-up additional signals. The "V" reflector concentrates all of the received energy on the driven element. The antenna is precisely aimed so that the inside of the "V" points to the television broadcast tower. Stations not in-line with the antenna are effectively ignored. The yagi corner reflector is a good, all-around antenna for receiving UHF stations, but it's not necessarily the best in all situations.

You wouldn't know it by looking at it, but a *fan dipole,* sometimes called a "bow-tie" antenna, coupled with a mesh reflector, is often very good at picking up UHF signals when the antenna is used within close range of the station. The drawback of the bow-tie antenna is that it is not very directional. Sometimes you need to aim the antenna directly at the station to avoid picking up other nearby channels. Many are used just for FM reception. Another antenna suitable for receiving UHF stations is the simple dipole with added mesh reflector. The reflector can be flat, or it can be a parabolic shape for increased directionality.

COMBINATION ANTENNAS

If you need to tune in all channels, you can save some money by purchasing a combination VHF/UHF antenna (you also can use the antenna for receiving FM stations). The typical combination antenna uses yagi dipole elements for the VHF channels, and a corner reflector for the UHF channels. A band-splitter is used inside the house to separate the VHF, UHF, and FM frequencies so they can be applied to the proper inputs on your TV and stereo.

USING THE RIGHT OFF-THE-AIR ANTENNA

Now that you know about the different types of antennas for receiving television signals, you can better judge which one is right for your TV. Keep the following issues in mind when selecting an antenna.

SIGNAL STRENGTH

The stronger the signal from the broadcast station, the further the signal will travel before it becomes too weak. UHF stations need higher wattages because their high frequency signals are lost more easily when passing through rough terrain or over long distances. You should not and cannot expect to receive all stations equally well, even if all of the broadcast towers are located on the same mountain top.

SIGNAL PROPAGATION

Like sound waves, radio waves grow weaker the further they travel. You can count on ideal reception if you live no further than about 20 miles from the transmitting station. You may not even need an outdoor antenna if you live within 10 or 15 miles of the broadcast towers.

You can pull in most TV stations within a 50 to 75 mile radius with the average outdoor antenna. Longer distances—like 75 to 150 miles—require a better antenna, and perhaps even an antenna amplifier to help boost the signal.

TERRAIN

Terrain plays a crucial role in reception, even if you live near the transmitting tower. You get the best reception if you are in "line-of-sight" with the broadcast tower. The signal will be considerably weaker if you are nestled in a valley or tucked behind a mountain where the signal doesn't easily reach you.

DISTANCE

If you live within 35 to 40 miles of the broadcast towers, you can probably get by with a small antenna, one with as few as 6 to 10 elements. Larger antennas—say 15 to 30 elements—are needed if you live further away, or if you are not in line-of-sight with the transmitter.

••• INSTALLING HOME THEATER •••

USING AN INDOOR OFF-THE-AIR ANTENNA

If you're lucky enough to live near television broadcast towers—say no more than 10 or 15 miles—you might not need an outdoor antenna at all. An indoor, table-top antenna could be all you need to bring the pictures into vivid focus and remove the rough edges from the sound. An indoor table-top antenna, such as that shown in *Figure 6-3,* may also be a necessity if you live in an apartment, mobile home, or dormitory, and cannot erect an outdoor antenna.

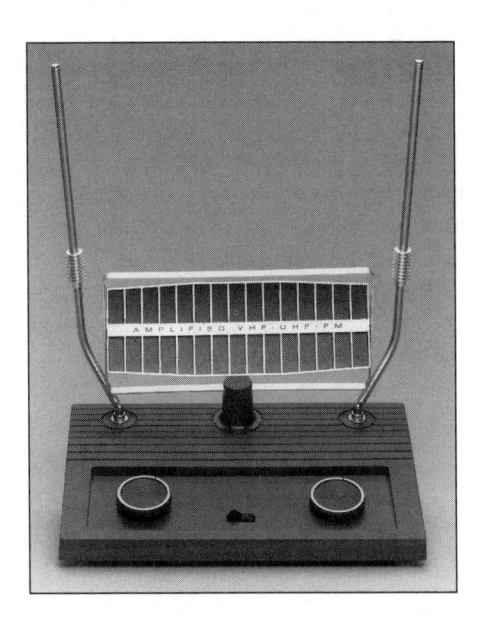

Figure 6-3. Indoor antennas are ideal if you live near the television broadcast towers. Select an antenna to meet your reception requirements.

Indoor antennas run the gamut from simple "rabbit ears" to more sophisticated consoles that rest on the top of your TV. Don't be fooled by appearance: some of the better indoor antennas—despite looking like props for a 1950s Buck Rogers movie—can substantially improve television reception. Many offer fine-tuning and phasing controls and let you position the elements for best reception. Indoor antennas are adversely affected by their immediate surroundings. They don't always work as well in all-metal mobile homes or trailers, and their reception is greatly restricted if your house is lined with aluminum-backed insulation. If you are in a shielded environment like this, placing the antenna near a window often helps improve reception.

There is another alternative to an indoor, tabletop antenna, you may wish to consider—installing an attic antenna. These are outdoor antennas specifically designed—gain, directionality, and size—to be used in an attic. In many cases, an ordinary outdoor antenna can be used if space allows. Check the specifications that come with the antenna to be sure.

Satellite Dish & "Off-the-Air" Antenna Hookups   6-9

••• INSTALLING HOME THEATER •••

ACCESSORIES FOR IMPROVING OFF-THE-AIR ANTENNA RECEPTION

You can't expect a cheaply made antenna to adequately pick up a station 150 miles away. But sometimes, even a good antenna doesn't pull in the signal as well as you would like. This is especially true during bad weather, which can adversely affect reception. Antennas are rated for clear weather only, with transmitting towers in line-of-sight. Of course, there's no where on earth that's perfectly flat *and* isn't plagued once in a while with bad weather, so use antenna ratings only as a guide.

Assuming that your antenna is at least minimally suited for the job, you can improve reception by adding a few accessories to your video gear:

- Install a rotator to aim the antenna directly at the broadcast station.
- Add an outdoor ("mast-mount") antenna amplifier, as shown in *Figure 6-4*, or an indoor amplifier. The outdoor amplifier attaches as near to the antenna as possible and boosts the signal immediately after it is received. Low voltage to power the outdoor amplifier is provided through the antenna lead. An outdoor amplifier is ideal if the antenna is some distance from the TV set, as it boosts the signal before its long journey from the antenna to the TV, overcoming signal loss in long cable runs. Indoor amplifiers attach near the TV, and should only be used if the incoming signal is already acceptable, not snowy. Otherwise, you merely amplify the snow, not the picture.
- Use top-quality lead-in wire, impedance matching transformers, signal splitters, band splitters, and the other paraphernalia associated with outdoor antenna installations (see *Figure 6-5*). Don't expect quality pictures with a $100 antenna if you use cheap or old lead-in wire.

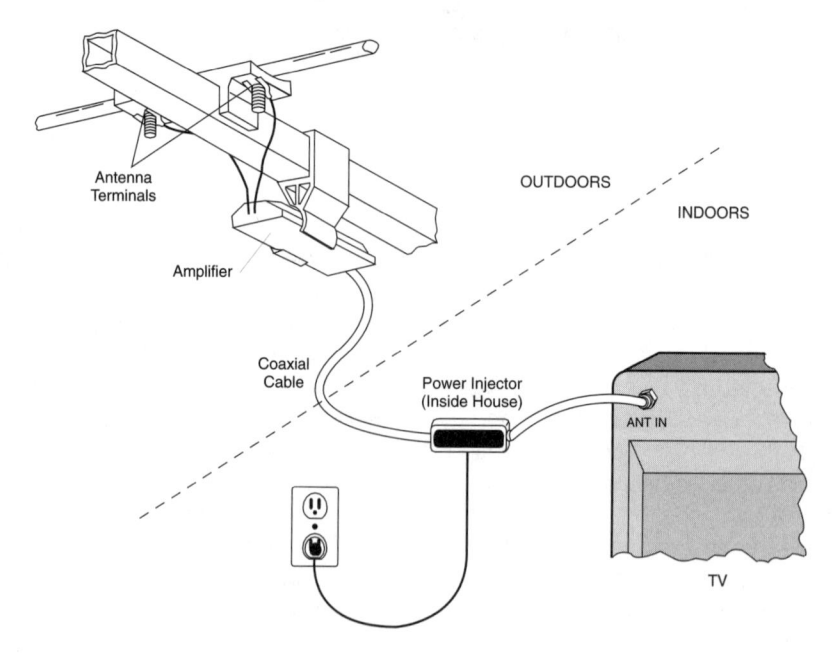

Figure 6-4. An outdoor amplifier consists of two parts: an amplifier module and a power supply. Power for the amplifier is provided through the coaxial signal cable. An outdoor antenna amplifier boosts signal strength immediately, before its journey through the coax to your TV or VCR.

••• INSTALLING HOME THEATER •••

Transformer Splitter
(RS #15-1139)

Signal Splitter
(RS #15-1293)

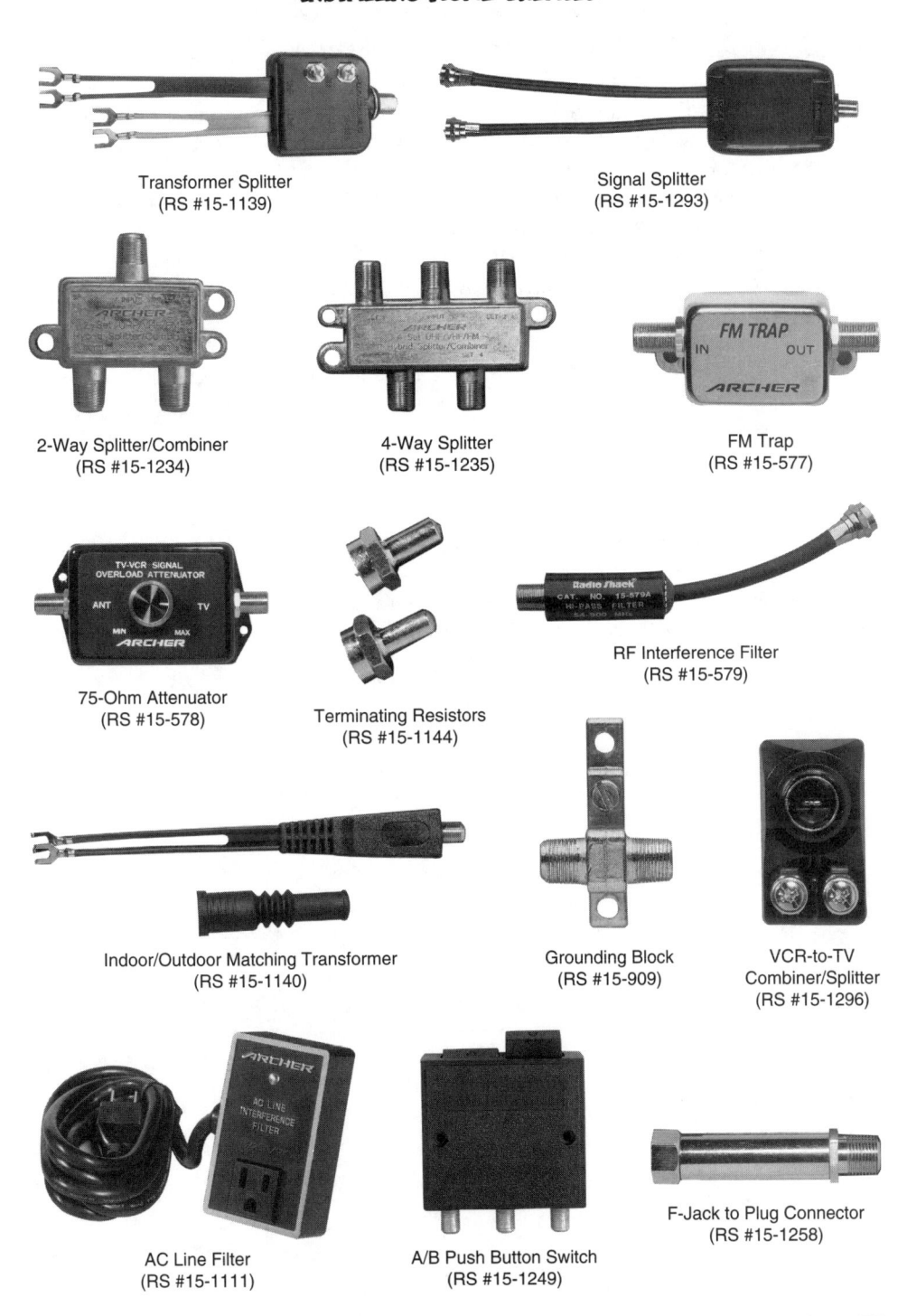

2-Way Splitter/Combiner
(RS #15-1234)

4-Way Splitter
(RS #15-1235)

FM Trap
(RS #15-577)

75-Ohm Attenuator
(RS #15-578)

Terminating Resistors
(RS #15-1144)

RF Interference Filter
(RS #15-579)

Indoor/Outdoor Matching Transformer
(RS #15-1140)

Grounding Block
(RS #15-909)

VCR-to-TV
Combiner/Splitter
(RS #15-1296)

AC Line Filter
(RS #15-1111)

A/B Push Button Switch
(RS #15-1249)

F-Jack to Plug Connector
(RS #15-1258)

Figure 6-5. A wide variety of accessories are available that can improve the performance of your TV antenna system, such as those shown here. The basic function of each accessory is outlined in the illustration. The RS number is the Radio Shack part number.

Satellite Dish & "Off-the-Air" Antenna Hookups

Where to Aim Your Off-the-Air Antenna

TV station broadcast towers are rarely, if ever, located next to their studios. Most often, the studio and offices are in or near downtown and the transmitter tower is perched on a nearby hill or mountain—or, in major cities, on top of skyscrapers. The higher elevation of the transmitter helps the signal reach more homes in a larger area.

Even if you know where a particular TV station is located, you may not know the location of its broadcast tower so that you can accurately point your antenna at it. But you can readily find out by calling the TV station and asking where its tower is located. *Sometimes,* you will get a sharper picture by aiming the antenna a few degrees right or left, rather than directly at, the tower. Experiment to find the best spot.

The task of locating the TV tower and aiming your antenna is a lot easier if the antenna is mounted on a rotator. Watch the set as you turn the rotator knob. Stop when the picture is the strongest. Note that the picture will come and go as the antenna turns; you may even get an acceptable picture with the antenna pointed 180 degrees in the opposite direction. But the reception will be best at one spot only.

Wiring The Complete System

Some home theater system installations may require fairly simple wiring. Others may require special wiring, such as that needed to provide invisible installation of speaker wires, especially for rear-channel speakers. In the next chapter, you'll learn how to wire your home theater system so that it looks like a professional installation.

Wiring Your Home Theater System

Next to buying your home theater system, the most enjoyable experience is using it. On the other end of the scale, for most people the least enjoyable experience is installing the home theater system in the first place! This includes wiring TVs, speakers, and other components to complete the home theater system.

Wiring a home theater can be a fun and rewarding experience, but you should know the pitfalls before you start. To do the job right, pre-plan your installation so you can tailor the wiring job to the design of your home. You'll want to hide the wiring inside the walls, along baseboards, or under the carpet. Depending on the layout and construction of your house and the room in which you are installing your home theater system, some alternatives detailed in this chapter will apply while others will not.

You also should be somewhat familiar with using woodworking tools. If you lack the skills and equipment, consider enlisting the help of a knowledgeable friend, or paying for professional installation. Working with electric saws and drills can be dangerous if you don't know what you are doing.

The Tools You'll Need

You can't wire your home theater system without some basic tools, many of which you probably already have in your tool kit. If you don't, consider buying them, borrowing from friends (if they still trust you to bring their tools back), or renting them.

BASIC TOOLS
- Claw hammer
- Screw driver assortment
- Pliers, small and large
- Utility knife
- Crimping tool for F-connectors
- Drill (manual or motorized; motorized recommended, rechargeable preferred)
- Drill bit assortment
- 8-inch long wood auger for drilling holes through walls

OPTIONAL TOOLS (DEPENDS ON INSTALLATION)
- Motorized saber saw
- Router and router bit assortment
- Volt-ohm (VOM) meter for testing cables and connections
- "Fish wire" for fishing cables through walls
- Density-type stud finder

••• INSTALLING HOME THEATER •••

PLANNING THE JOB

Before drilling a single hole, map out *exactly* what you plan to do before the work begins. Start with a pencil sketch of the layout of your home. This floor plan doesn't have to be to scale, but the layout should be as accurate as possible. Using graph paper for the layout accomplishes both. Mark where you want the wiring to go and how it will get there. Take into account the fact that you may have to go over or around doorways and windows. Also, consider what is above (2nd floor bedroom or attic) and below (basement, crawl space, or nothing) your home theater room.

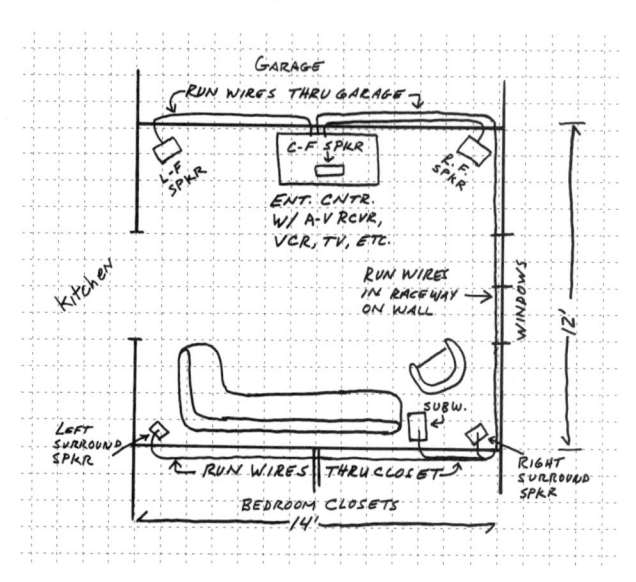

Figure 7-1. Use graph paper to draw a layout for your home theater room. It will help you plan and understand everything you need to do to complete your installation.

The bulk of the wiring tasks to be accomplished fall into four categories:
- Cables connecting the TV, VCR, A/V receiver, and other components.
- Wires connecting the A/V receiver to the speakers.
- Lead-in cables from an outdoor antenna (including satellite dish) to the VCR or TV.
- Telephone line for ordering pay-per-view programming.

Your floor plan will help you calculate the approximate length of wiring and cabling you will need, as well as any extra parts, such as coax F-connectors, wall outlet plates, through-wall tubes, telephone jacks, and so forth. Make a shopping list and get everything you need at once, before you begin, to eliminate return trips to the store. And, with the requirements of the job drawn on a floor plan and a complete shopping list, you can more easily determine if the scope of the installation is beyond your skills, endurance, or budget.

It's also a good idea to find out exactly what you can and can't do with the home theater system in your home. Many cable companies prohibit you from tapping into their line and adding an extension—even if the cable is on your property and in your house. These restrictions come in the form of local ordinances, so you can check on the legality of splicing into the cable by calling your cable company.

••• INSTALLING HOME THEATER •••

FOR STARTERS: UNTANGLE THE MESS BEHIND THE TV

Before you start any wiring job, be sure that the cables and wires behind your TV connecting the A/V receiver, VCR, and other components are neat and organized. Jumbled wires can cause signal degradation as well as add to your frustration in case things don't work correctly on the first try. If you haven't already done so, inspect the wiring and look for the following warning signs:

- *Extra long lengths of coax.* Cut these to length and attach a replacement F-connector at the end of the cut piece. F-connectors and crimping tools are available at Radio Shack.
- *Kinks in coax.* Coax cable is designed so that the center conductor is kept a certain distance from the outside conductor. A kink changes this distance and can alter the way the sound and pictures are transmitted through the wire. Remove the kink or replace the cable if it's damaged.
- *Coils of video and audio cable.* Coiled wires form a type of antenna. Avoid looping any of the connecting cables or power cords.
- *Power cables bundled with video and audio cables.* Power cords carry a constantly pulsating ac electric field that can be electrically induced into nearby signal cables, causing interference. Always try to physically separate power cords from audio/video signal wires. Even a distance of a few inches can eliminate interference.
- *Broken or dirty connectors.* Replace the connectors (or the entire cable if necessary) if they appear broken (see *Figure 7-2*). Clean connectors with a suitable cleaner, or use isopropyl alcohol
- *The wrong wire.* Coax cable is only suitable for carrying RF television signals. If your system uses the older, twin-lead cable, replace it. Baseband (non-RF) audio and video signals should be carried through shielded cable only. While you can easily go overboard with expensive cabling, the idea is to get the proper type for the job. Gold plating, pure silver conductors, or fancy rubber insulation are optional. The rule of thumb is: get the best cable you can afford, but don't go broke in the process.

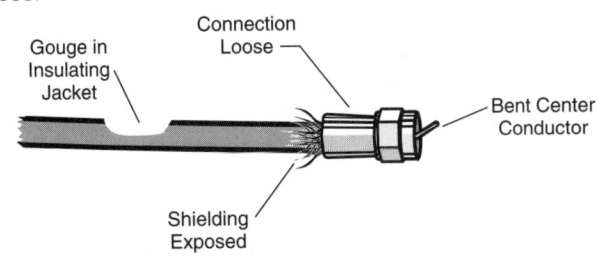

Figure 7-2 Broken connectors and cables can seriously degrade the performance of your home theater system. Repair, replace, or clean the cabling and connectors as necessary.

For ease of installation, clearly mark each end of every cable as you connect your record changer, cassette deck, CD player, DVD player, TV, and antenna to the A/V unit. Cable marking kits are available at Radio Shack, or you can use masking or white tape (use dark pencil or felt-tip; ball-point pen ink can smear). And, again, be particularly wary of power cords that are intertwined with audio signal cables. This can cause an annoying hum that can be very difficult to locate and eliminate. You may wish to use split tubing, available at Radio Shack, to keep power cords separated from cables and wiring. To use, you merely slip the wiring into the tubing. The tubing is available in various diameters.

Wiring Audio and Video Through Your Home

Connecting a wire between an A/V receiver and a speaker is a fairly simple and straight-forward. The larger issue is where to put the wiring so that it is hidden from view once the system has been installed. Wires leading to and from your home theater system should be out of the way, both visually and physically. Speaker cables strewn carelessly along the floor can create a tripping hazard that may cause an injury to you or to your expensive home theater equipment.

Depending on the construction of your home, and whether you are interconnecting units that are internal or external, you have several alternatives for hiding the wires (in the following discussion, the word "cable" is used as a generic term for video and audio cable, speaker wires, telephone wires, and coaxial cable):

- *Run the cable through the basement or crawl space.* This is probably the easiest method, and it's nearly invisible. For an expert job, you'll want to fish the cable through the wall to the basement.
- *Run the cable through the attic.* If your house doesn't have a basement or crawl space, this is the most desirable way to run the cables, even though it can be more difficult.
- *Stretch the cable along the baseboard.* In houses that lack basements and attics, this is an acceptable alternative, even though it may involve more work if you want to hide the cable.
- *Place the cable under the wall-to-wall carpeting.* This approach is acceptable only when the cable is not under a high-traffic area. This usually means running the cable along the edges of the room.
- *Run the cable outside.* Use this method for bringing antenna signals from the outside into the house. Make sure only to use coaxial cable rated for exterior use. Use coax retainers (these nail into wood or plaster) to keep the cable in place. If you wish, paint the cable the same color as the house.

Let's examine each method more closely, and describe the steps to do the job.

Basement Cabling

If your home has a basement, or even a crawl space, consider using it for home theater wiring. You can run most any type of wiring through the basement.

If you seldom venture into the basement (especially if it's unfinished) or crawl space, think about fumigating it first. That will reduce the possibility of being bitten or stung by insects that have taken up residence under your house. Always wear protective clothing, including long sleeves and gloves, if possible.

Figure 7-3 shows a typical basement wiring project. Note that the wire is first inserted into the wall, and that a hole is drilled in the toe plate (the piece of wood—usually a 2 x 4—that forms the bottom of the wall) for the cable to pass through. You will need to drill up from the basement to make this hole. You can reliably determine where to drill if you first drill a "pilot hole" at the corner of the room, as shown. Insert a wire or coat hanger into the hole. Use this wire as a guide when drilling in the basement. Remove the wire, and patch the pilot hole, if necessary.

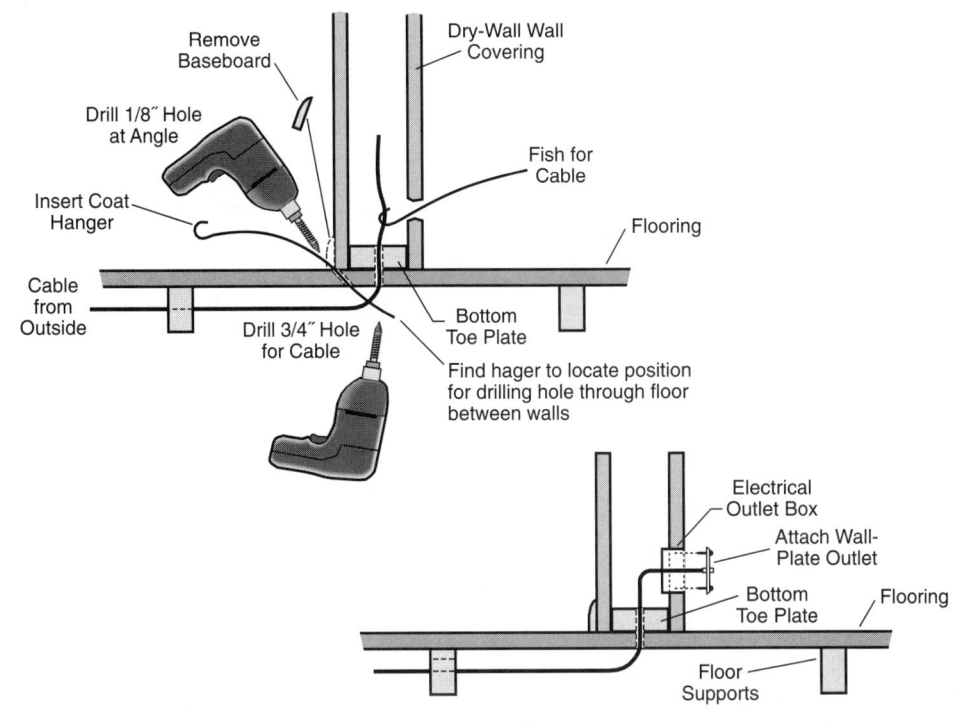

Figure 7-3. When using the basement for routing home theater cabling, drill a small pilot hole, as shown, to help you locate where to drill while in the basement. The cable/wire can be routed through the hole in the floor and toe-plate, and inside the wall of the room.

Enlist a friend to help fish the wire through the wall. Use a fish wire, coat hanger, or other flexible rod and insert it through the hole in the toe plate and into the wall. Have a person on the other end attach the wire to the rod (use strong tape), then carefully pull the rod through. Repeat the process at the other end. For best results, buy a length of "fish wire" at the local hardware store. This stuff is specifically made to route wires through walls.

ATTIC CABLING

More homes have attics — or "attic-ish" crawl spaces — than basements. Attics are an alternative route for the wire if your house lacks a basement, or if the basement isn't suitable for TV wiring.

Fishing wires through the attic is more difficult, however, if you want the job to be invisible. You must first route the wire up the wall and into the attic, which can be extremely difficult, if not impossible. The reasons: the distance between the hole and top plate of the wall can be in excess of six feet (conversely, it's less than two feet when routing a cable through the basement), and attics lack head room making it more difficult to work. That makes it hard to fish the wires through the wall. In fact, cutting a long hole in the wall and patching it later often is easier! Plus, if the wall faces the outside of the house, it will likely contain a horizontal stud at waist level, as shown in *Figure 7-4*. This serves as a fire break, and in some states (such as California) is required for seismic codes. In many newer homes, this horizontal waist-level stud is found in all walls, exterior and interior.

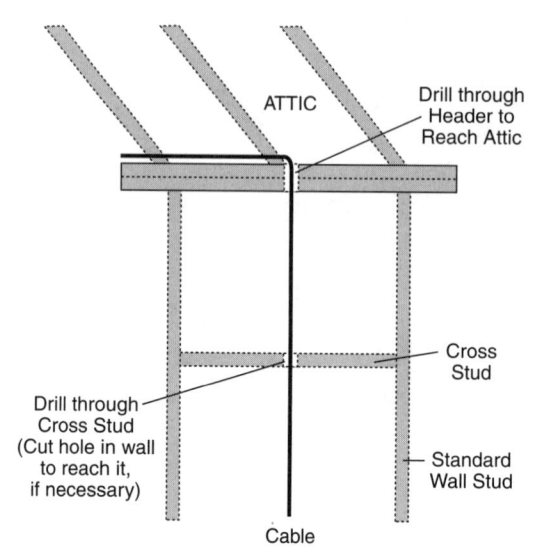

Figure 7-4. Attic cabling is generally more difficult than basement cabling because of the extra distance involved, low head room when working in the attic, as well as studs and other obstacles commonly found inside walls. You must route the wiring around these obstacles to reach the attic. In some cases, this involves making fairly large holes in the wall, then patching later.

If you're not sure of the construction of the walls in your home, buy a "stud finder" at the hardware store. Be sure it's the kind that registers changes in density, *not* the metal of nails. Stud finders that rely on locating the framing nails are less expensive but are not as reliable.

There is an alternative if you don't mind a little wire showing. You can carefully route the cable from the attic down a corner of a wall or in the a corner of a closet. The wire travels from the home theater unit along the baseboard to the corner of the wall. The wire is taped or cemented into the corner as it travels up the wall. A hole is drilled through the top plate or ceiling to lead the wire to the attic. For best aesthetics, paint the wire the same color as the wall.

DEALING WITH A MULTI-STORY HOUSE

Homes with more than one story present special problems for installing a home theater system. The biggest problem: getting the wire up the wall and into the next floor. As in attic installations, it can be extremely difficult to fish wires through a wall and into the floor above—so much so that most professional installers don't even attempt it. Instead, they look for alternatives:

- Routing the wire outside, where it's easier to traverse the wall.
- Finding a nearby closet or wall storage space and install the wires inside it.

••• INSTALLING HOME THEATER •••

- Cutting a long hole in the wall and placing the cable(s) inside it. A piece of drywall is used to patch the hole and then spackled or plastered and painted to match the rest of the wall.
- Routing the wire up a corner of the wall and painting the wire to help hide it.

If you have wall-to-wall carpeting on the second floor, you can pull it back to expose the hole you need to drill through the ceiling. Route the wire along the baseboard or tackless strip, as desired, to bring it to its destination.

BASEBOARD AND UNDER-CARPET WIRING

Generally considered a last resort, you can install home theater wiring along the baseboard of rooms, and even under the carpet if you have wall-to-wall carpeting. Both methods require special care, or damage to the cabling and carpeting could result.

For baseboard installation: If the wire is small (speaker wire, for example), you can sometimes tack it directly to the baseboard, as shown in *Figure 7-7a.* You can use staples, but only the U-shaped kind with insulation; these are designed for wiring purposes. Otherwise, the staple could cut through the insulation of the wire and your entire home theater system might be plagued by short circuits. A better method is to remove the baseboard and cut a groove or notch inside to accept the cable. Use a table saw, radial arm saw, or router to cut this groove. Make the groove or notch just large enough for the wire, but be careful not to remove so much wood that the baseboard is weakened. For best results, remove the old nails from the baseboard and fill the nail holes with wood putty. Re-stain or re-paint the baseboard. When dry, insert the wire into the baseboard and attach the baseboard to the wall (be sure that you don't nail through the wire!) using 2d or 3d finishing nails. By using new nails and nail holes, the baseboard is not as likely to come loose.

For wall-to-wall carpeting: As shown in *Figure 7-5a,* use a heavy pair of pliers to pull back a corner of the carpeting. Place the wire between the tackless strip and the padding—*never* between the wall and the tackless strip; that space is required to properly install the carpeting. As shown in *Figure 7-5b,* if necessary cut out a small strip of padding if the wire won't comfortably fit. Replace the carpeting after the cable is in place. Use a rubber mallet and block of wood to gently re-tamp the carpet back onto the tackless strip. Avoid bending the "tacks" in the tackless strip, or the carpet won't stay down.

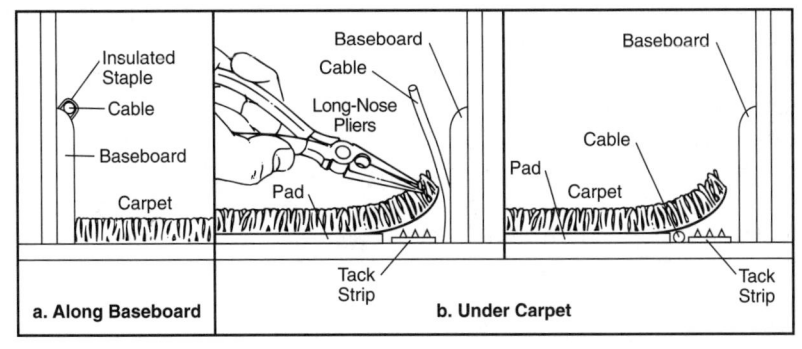

Figure 7-5. a. You can tack speaker wire directly to the baseboard using insulated staples. b. Place the cable/wire between the tackless strip and the padding. Be sure to firmly re-attach the carpet to the tackless strip. Don't pull up too much carpeting at once, or it will be difficult to put it back.

Avoid removing carpet thresholds around doors and tile floors. Instead, pull up the carpet to either side, and slip the cable underneath. You can also stretch the wire across the carpeting, and not just along the baseboard. However, you'll need to remove more of the carpet from its tackless strip, and fold enough of it back to reveal the part of the floor you need access to. You should also cut a notch in the underlying pad and press the cable into the notch. This eliminates the "hump" that can appear when the carpet is replaced, and helps prevent pressure damage to the cable.

CAUTION: It is generally a bad idea to completely uproot a wall-to-wall carpet. Try to take small portions at a time and replace the carpet before continuing with the next section. Most wall-to-wall carpets are installed under tension;; that is, they were cut a little short and stretched to fit. This prevents wrinkles and sagging. If you pull up too much of the carpet, you may have difficulty returning it to its original size and condition. Should this happen, call in a professional carpet layer who will use various "kicker" and stretching tools to bring the carpet back into shape.

Outside Cabling

Routing wire outside is ideal if you're adding a cable TV or antenna extension to other parts of the house. Merely tap into the main cable or antenna lead before it enters the house and use a splitter (see Chapter 6) to divert the signal. You can install the cable close to the ground or under the eaves. For cable TV extensions, the ground approach is more desirable because it will be largely hidden from view. You won't have to deal with a wire stretching from the roof, down the wall, and into a hole that leads to the bedroom. You can use either method for TV antenna installations, as the wire has to travel from roof to ground level at some point.

Keep in mind that the coax cable you use should be rated for outdoor use. Not all coax cable can withstand the elements, particularly in varied climates that see very cold winters and hot summers. Using the wrong kind of cable will cause it to deteriorate faster, requiring you to replace it after just a few years of use. Any coax cable suitable for use with an outdoor antenna is acceptable for use in outside cabling. If you have particularly long cable runs, use RG-6 coax because it has less signal loss.

When splitting the incoming cable, remember that signal strength is affected. In a two-way splitter, one half of the signal strength will go down one line; the other half down the opposite line. Splitting the signal more than two ways is generally not recommended, unless you use an amplified splitter. These typically require indoor installation.

Figure 7-6 shows how to route an outside extension cable. *Figure 7-6a* is a side cross-sectional view; *Figure 7-6b* is an outside front view. Holes must be drilled in the wall where the cable feeds through. Note the use of through-wall tubing, which makes for a longer-lasting and more professional installation. After feeding the cable through the tube, seal it with silicone sealant or caulk. Use suitable cable staples, as shown, to anchor the coax to the wall. Consider painting the cable the color of the house. Make sure to put a "drip loop" on the cable before it enters the house, as shown in *Figure 7-6*. This simple loop allows rain water to run down to the bottom of the loop and drip off, rather than run into the house along the cable.

••• INSTALLING HOME THEATER •••

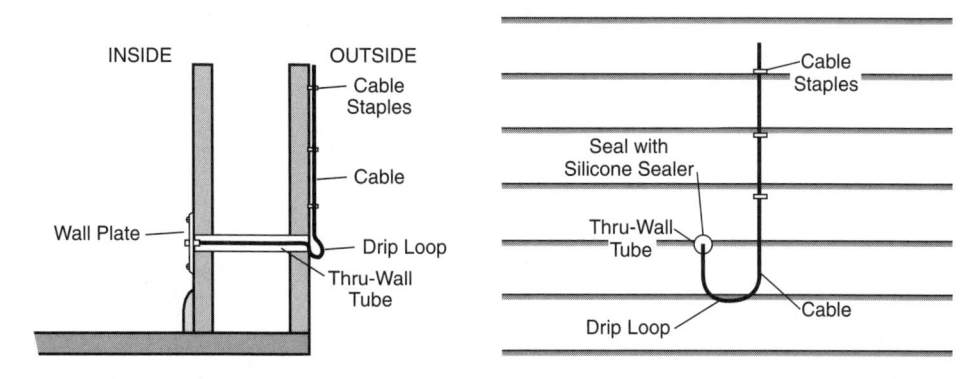

Figure 7-6. Use through-wall tubing when routing a cable from the outside to the inside. Staples help secure the cable to the side of the house. Make sure the put a "drip loop" in the cable line just before it enters the house. This helps keep rain water from entering the house when it runs down the cable. Inside, the cable can terminate at a wall face-plate, or it can extend to another part of the room.

MOUNTING SPEAKERS

Most speakers (either main or surround) for a home theater system can be placed on shelves or tables, or they can be hung from a wall or ceiling. When hanging speakers, you'll want to use the proper brackets. Make sure the brackets are rated for the weight of the speakers. The brackets should be attached to a stud inside the wall; if the speakers are lightweight—as most surround speakers are—you can mount the brackets to a wall using expanding anchors.

Don't mount a subwoofer on a wall, or even place it on a book shelf. These speakers provide the very deep bass sounds and create a lot of vibration. The vibration of a subwoofer mounted on a wall could actually cause the wall brackets to come loose. Place a subwoofer on the floor, out of the way of traffic and even out of sight.

HIDING SPEAKER WIRES

Mounting the speakers is only half the battle; hiding the wires is another. If there are no horizontal fire-block studs used in the wall, you may be able to fish the wire through from the floor or basement. This requires some diligence and patience, however, as the task is not as simple as it looks. Another way is to consider what's on the other side of the wall from where the speakers are to be mounted. If it's a closet, garage, or storage area, you can pass the wire through to the other side of the wall. Drill holes at the bottom of the wall, feed the speaker through, run it up to the proper location and height, and then feed it back through to the speakers. You can also carefully thread the speaker wires up the wall or along the corner of two adjacent walls. Attach the wire securely to the wall, and paint the wire the same color as the walls.

USING RACEWAYS

For a utilitarian look, use a plastic wiring raceway, available at many hardware or electrical supply stores, as shown in *Figure 7-7*. These are primarily intended to add ac electrical service inside rooms where in-wall wiring is not possible (along a brick wall, for instance), but they can also be used with speaker wire. The raceway consists of a

Wiring Your Home Theater System   7-9

U-shaped plastic channel and connecting pieces. The channel is attached the wall, is available in a variety of colors, and can be painted.

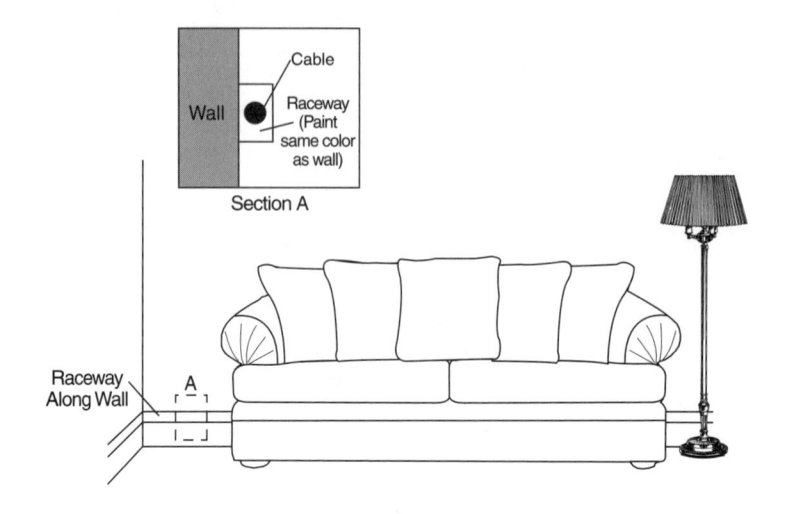

Figure 7-7. Plastic raceway can be used to help hide cable or wiring along a wall. In the case of brick or masonry walls, it is sometimes the only alternative you can use. You may paint the raceway to blend in with the rest of the room.

INSTALLING A CABLE OUTLET PLATE

If you are installing new coaxial cable for a satellite dish or an outside antenna, you may wish to install a cable outlet plate rather than just fish the wire through a hole in the wall. (Note: use the same general concept outlined here for telephone outlets. You may need a phone outlet near your home theater if your satellite or cable service offers pay-per-view programming. Read *Installing Telephones,* available at Radio Shack (RS #62-1060) for information on adding a telephone extension.) The cable you've installed terminates at the inside of the plate. You merely hook up your components to the connector on the front of the plate.

There are two ways to install a cable outlet plate: the easy way, and the hard way. Both look about the same on the outside, but the hard way offers a more permanent installation.

The easy way is to cut a hole in the wall large enough for the end of the cable. Pull the cable through and install a male F-connector. Screw the cable to the inside of the wall plate, then attach the plate to the wall. Depending on the composition of the wall, you may need to use plastic anchors, expanding bolts, or masonry bolts (for brick). This is shown in *Figure 7-8a.*

A more difficult, but permanent, approach is to cut a hole in the wall large enough to mount an electrical outlet box inside the wall, as shown in *Figure 7-8b.* The hole should be cut next to a stud, which means you'll have to use a stud finder to find the right spot. You'll need a motorized jigsaw to cut the hole. You can also use a keyhole saw but the work will go much slower.

••• INSTALLING HOME THEATER •••

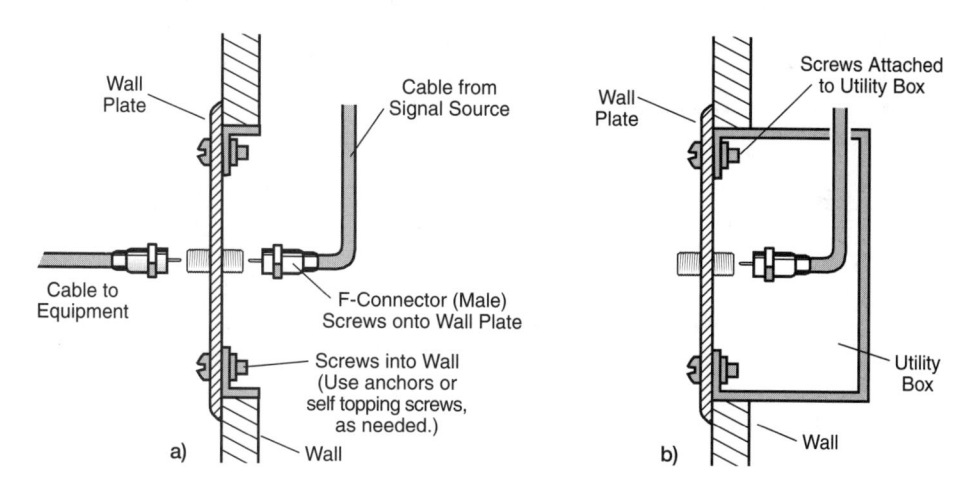

Figure 7-8. a) Attach the cable face plate to the wall using suitable screws or anchors. b) Install a utility box in the wall, and attach the cable face plate to the box. The utility box provides for a stronger, more permanent installation.

Be careful to make the hole just large enough for the box. Secure it to the side of the stud using screws (preferred) or nails. Feed the cable through the box, attach an F-connector, and secure it to the inside of the cable wall plate. Put the plate over the outlet box, and secure it into place using the screws provided. Electrical outlet boxes and cable outlet plates are available at most hardware and home improvement stores. The outlet plates also can be purchased at Radio Shack.

Do's AND Don'ts FOR BETTER INSTALLATIONS

In review, be sure to consider these important reminders about the do's and don'ts for better home theater installations.

- *Do* use the proper coax cable for all RF video signals. Specifically, use RG-59 or RG-6 cable, which is designed for video.
- *Do* keep wires as short as possible. Cut audio and video cables to length, and add your own connectors. The connector ends and suitable crimping tools (no solder required) are available at Radio Shack.
- *Do* separate signal wires from power cords. This reduces hum and picture interference.
- *Do* cut a notch at the edge of a carpet pad so you can fit the wiring inside. Otherwise the carpet may not lay flat.
- *Do* use speaker mounting brackets that are large enough and strong enough to hold the speakers you are using. Mount the brackets to the wall using a suitable anchor or, better yet, screw the bracket directly into the stud inside the wall.
- *Do* use silicone or other weatherproof sealant to fill the hole when passing a wire through an outside wall. This keeps out cold air, rain, and—yechhh—bugs.
- *Don't* attach wires to walls with staples, unless the staples are meant for cabling and have rubber insulation.
- *Don't* use twin-lead cable for carrying RF video signals, either inside or outside the house. It doesn't offer the same interference rejection as coax cable.

Wiring Your Home Theater System

- *Don't* coil excess wires. This can cause interference.
- *Don't* stuff wire between the wall and tackless strip used to hold down wall-to-wall carpeting. The carpet won't hold as well to the strip. Instead, temporarily lift the carpet and place the wire between the strip and the pad.

Considering The Room

This chapter detailed how to wire your house for home theater. It included a discussion of the tools needed for the job, as well as several approaches to installing hidden wiring. It also provided some helpful hints on creating a better environment for your home theater cables and wires by cleaning up any tangled mess that might be behind your TV and audio gear. In the next chapter, you'll learn about room design and ergonomics for home theater, including the best overall layout for home theater components, as well as the best placement for speakers and lighting.

ROOM DESIGN AND ERGONOMICS

You may have the latest whiz-bang projection TV, the best digital video disc player on the market, the most stupendous A/V receiver, and the finest speaker system ever made. But none of it matters if your high-priced gear is used in a room that is not suitable as a home theater auditorium. Simply stuffing home theater components into a living room or den doesn't make it a home theater—you must consider the size and shape of the room, the seating arrangement, lighting conditions, and more.

No, this doesn't mean you have to hire an interior designer to custom craft a room just for your home theater system. But you should take the time to consider the best location for your TV, and the size, shape and acoustics of the room, and whether or not you will need to use any sound-deadening materials to help soak up excessive echo.

In this chapter, you'll learn how sound plays a critically-important role in the design of a good home theater den, how sound is absorbed or reflected in a room, how to deal with light and glare, tips on placing speakers, the best location for the audience, and more.

SOUND PLAYS AN IMPORTANT ROLE

In order to fully appreciate the importance that sound plays in the quality of your home theater, it is necessary to understand the basics of the science of acoustics. You certainly don't need to be an sound expert to install, use, or maintain a home theater system. But you will derive greater pleasure from the system if you understand the basics of what makes up sound.

WHAT MAKES UP "SOUND"

On a broad level, the sound you hear—from a home theater system or any source—is the movement of air molecules toward and away from you. Sound starts at a source, such as a person's mouth, or the speaker of a home theater system. The sound causes the surrounding air to vibrate. This vibration is in a specific pattern called a wave, and this wave is similar to ripples on a pond when a pebble is thrown into calm water.

Sound waves cause the air to alternately compress and expand. That is, at one instant the air molecules may be pressed tightly together, and at another instant, the air molecules may be spreading apart. This change of pressure—compression and expansion—is what actually makes up sound, and causes our eardrums to vibrate when the sound wave reaches us. The ear drum is pushed in on the expansion phase of the wave, and pulled out on the compression phase (see *Figure 8-1*).

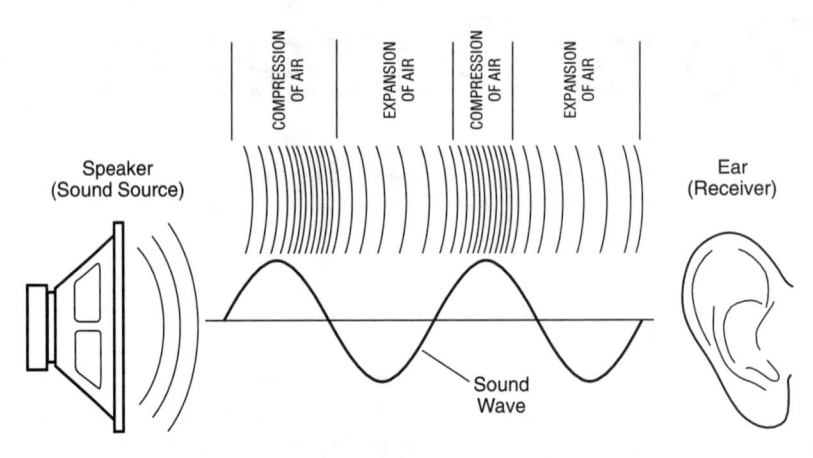

Figure 8-1. Sound is passed through air by alternately compressing and expanding pressure. Our eardrums move in response to the pressure changes, and this movement is transmitted to the brain as sound.

THE CONCEPT OF SOUND VOLUME

The amount of air molecules affected by the sound wave determines the loudness of the sound. Speak in a whisper and only a small amount of air vibrates; the volume is said to be "low." Shout at the top of your lungs and a lot of air vibrates; the volume is said to be "high." *Figure 8-2* shows a visual depiction of differences in sound volume.

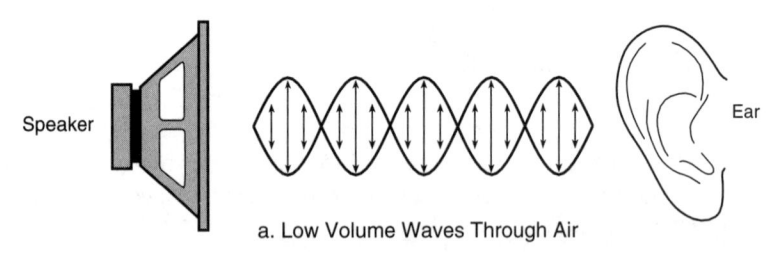

a. Low Volume Waves Through Air

Double arrow indicates up and down motion of air particles that propagate the sound wave from source to receiver. The particles remain in place; the wave propagates. Larger amplitude indicates louder sound.

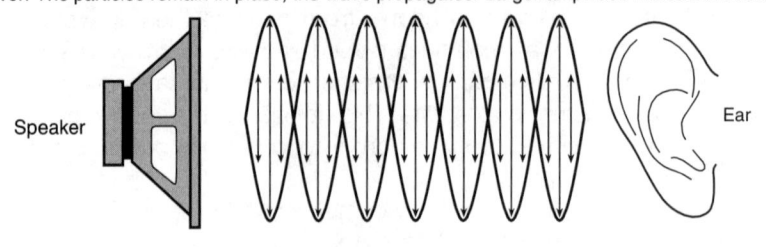

b. High Volume Waves Through Air

Figure 8-2. The changing pressure of a small amount of air results in quiet sounds. Conversely, the changing pressure of a large amount of air results in louder sounds.

Room Design and Ergonomics

••• INSTALLING HOME THEATER •••

Volume is actually a poor term to use when defining the loudness of sound because the term is inexact. While "volume" is certainly an acceptable word, a better term that describes the loudness of sound is *sound pressure level,* or *SPL.* SPL is commonly expressed in decibels. The decibel (dB) is not an absolute measure, but rather a ratio between two sound levels. Zero dB represents the SPL of a test tone that can just be heard by a person with normal hearing. The level of other sounds is measured or expressed relative to this lowest audible sound level. For a person of normal hearing, a change in SPL of 10 dB is perceived as being twice as loud as the original sound.

THE CONCEPT OF SOUND FREQUENCY

Try this simple experiment the next time you're at a lake, pond, or pool. With the water calm and smooth, take two rocks, a small one and a large one. Drop the small rock over one side of the boat. Now drop the big rock over the other side of the boat. The waves from the small rock are closely spaced, whereas the waves from the big rock are spaced further apart.

The spacing of the waves determines its frequency. Frequency is a measure of how many waves go past a stationary object in one second. Waves spaced far apart have a low frequency, while waves spaced close together have a high frequency. In sound, the frequency determines if the sound is high pitched or low pitched. As you might surmise, a low-pitched sound has a low frequency, and a high-pitched sound has a higher frequency. Sound is expressed in Hertz (Hz). One Hertz is equal to one sound wave cycle in one second. For example, 100 Hz is 100 sound wave cycles per second.

Humans hear in a relatively narrow range of frequencies, from about 20 Hz to 20,000 Hz (also expressed as 20 kHz; the k represents thousands). Sounds above and below these frequencies are considered subsonic or supersonic, and are outside the range of normal hearing. The ability to hear very low and very high frequencies depends on age, sex, and exposure to loud sounds. Generally speaking, most men past age 50 have difficulty hearing sounds above 15,000 Hz, whereas many teenagers (male or female) are able to hear sounds in excess of 20,000 Hz. While hearing varies from person to person, in home theater systems it is generally accepted that the "normal" range is 20 Hz to 20 kHz.

ROOM ACOUSTICS: SOUND IN AN ENCLOSED SPACE

The sound you hear is greatly affected by its environment. This is most obvious in a small, enclosed area with smooth, hard-surfaced walls. The sound of footsteps is completely different out in the open than it is in an empty school gymnasium, for example. In the open, you hear just the actual sounds of your feet striking the ground. In the gym, you hear the sounds of the footsteps plus numerous echoes as the sound bounces from wall to wall, ceiling to floor.

Most sound systems are used indoors and the size, shape, and other factors of the room greatly determine the quality of sound your home theater system will produce. There are two major ways enclosed rooms affect sound: reverberation, and its converse, absorption. The following discussion introduces the concepts of reverberation and absorption and their causes. Later in this chapter, you will learn how to control reverberation and absorption to achieve better sound.

REVERBERATION

Just like the school gym, the walls in a room will cause some echoes of the original sound. These echoes, which are more accurately called *reverberation,* may be insignificant. Or the reverberation may be so overpowering that you hear more echo than original sound. Contrary to popular belief, the "ideal" room is not reverberation-free. In fact, our ears and brain are trained for a certain amount of reverberation, so a completely echo-free sound may seem unnatural. Reverberation is part of normal sound and only excessive reverberation is considered detrimental to a sound system.

Reverberation is caused by sound bouncing off the boundaries of a room. These boundaries include walls, floor, ceiling, and any other object that is large and hard enough to reflect sound. Notice the word "hard." Soft materials, regardless of their size, cause little sound reflection. Rather than bouncing off the boundary, the sound is absorbed. More about sound absorption in a bit.

By their nature, reflected sounds are not as loud as the direct sound from a source, such as a speaker in a home theater system. The relative loudness of the reflected sound depends on a number of factors: how much sound is reflected from the boundary, the angle of the boundary in relation to the sound source and the listener, and whether the source of the reflected sound is itself a reflection. Look at a simple, square room (see *Figure 8-3*). The main sound source in the center of the room is reflected many times, often with reflections causing additional reflections. Remember, while these reflections may appear to make a muddle of the sound, as long as the reverberation is not overpowering, the echoes will be perceived by the human ear and brain as natural and acceptable.

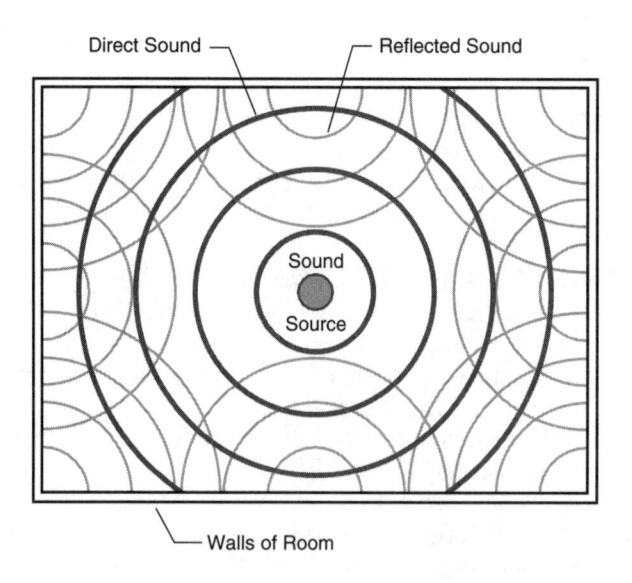

Figure 8-3. Reverberation causes a sound to echo. Most reverberation is caused by sound bouncing off walls, ceilings, and floors. However, any solid object larger than one square foot or so can cause sound reflections.

ABSORPTION

In a typical room filled with an audience, the bulk of the sound is absorbed rather than reflected. Recall that the *hardness* of the boundary largely determines how much of the sound is reflected. Hard boundaries, like a brick or smooth plaster wall, absorb very little sound—most of the sound is reflected, and is heard as reverberation.

Conversely soft boundaries, like carpeting, acoustic tile, drapes, soft furniture, and even people, can absorb quite a bit of sound. Because more sound is absorbed, less sound is reflected. This reduces the amount of reverberation. *Table 8-1* shows the relative absorption coefficients of different kinds of common boundaries. The absorption coefficient starts at 0, which means total reflection, and goes to 1, which means total absorption. Values in between are shown with a decimal point, making it easy to calculate percentage. For example, acoustic tile has an absorption coefficient of .70, or 70%

Material	Absorption Coefficient	Material	Absorption Coefficient
Acoustic tile	.70	Plaster wall	.06
Audience	.95	Plywood paneling	.20
Brick wall, painted	.04	Poured concrete	.02
Brick wall, unpainted	.02	Upholstered seats	.90
Carpeting	.30	Wood flooring	.08
Drapes	.75		

Table 8-1. Absorption coefficients of selected boundary materials.

Table 8-1 lists the absorption coefficient for a representative test sound at a frequency of 1 kHz. In most cases, absorption of a given material decreases with lower frequencies and increases with higher frequencies. The high frequencies are the first ones to be absorbed, leaving predominately the lower frequencies. This is why the sound may be overly "boomy" or "bassy" in some rooms.

THE EFFECT OF ROOM SIZE AND SHAPE ON REVERBERATION

The smaller the space, the more the sound will reverberate within that space. Large, open rooms tend to reverberate less because the sound is lost as it bounces from wall to wall. Open areas of the room, such as doorways to other parts of the house, also will have an effect on reverberation, allowing sound to escape the room. Only a relatively small portion of that sound is likely to re-enter the room. The height of the ceiling also plays an important role in reverberation. The higher the ceiling, the less reverberation.

Let's consider two extremes of home theater rooms and the reverberation that is likely to result from each:

Our first room is located in a basement and measures 8 by 12 feet. It has few windows and no drapes. The ceiling height is approximately seven feet. Two walls are white painted brick, and the other two walls are plywood paneling. The cement floor is covered with an area rug. This room will be very "lively" in terms of reverberation. The hard surface walls readily bounce the sound from boundary to boundary. The relatively small area of the basement room (width x length x height) enhances the reverberation. There are no window drapes or major open areas to other rooms to either absorb or allow some sound to escape. More than likely, this room will produce excessive reverberation and you will want to add fabric to some of the walls to help absorb the excess reverberation (see "Using Sound Absorbing Fabrics," below, for more information). The owner also may want to consider thicker carpeting, and perhaps adding a layer of acoustic tile to the ceiling.

Our second home theater room is a spacious family room or "great room" measuring 12 by 18 feet. This room adjoins an open dining area and kitchen. There are only the two outside walls at right angles to one another; the rest of the family room is open to other parts of the house. The room has a vaulted, 12-foot ceiling, wall-to-wall carpeting, and drapes covering the two large picture windows. The walls are made of gypsum wallboard. All of these materials absorb sound. Such a room is likely to produce very little reverberation due to its large size, open architecture, and the damping characteristics of the carpeting and drapes. In fact, this room may exhibit too little reverberation, requiring boosting the sound level on the amplifier to compensate.

Most rooms are somewhere in between these two extremes, but the effects of reverberation of these two types of "troublesome" rooms illustrates the issues involved. As you are planning your home theater room, consider:

- The size of the room. Medium sized rooms (between about 120 and 300 square feet) are ideal. Reverberation—particularly of low frequencies—tends to be a problem with small rooms.
- The height of the ceiling. Low ceilings accentuate reverberation, while high ceilings attenuate reverberation.
- Floor covering. Hard floors increase reverberation. One of the simplest ways of controlling reverberation is to cover the floor with carpeting. Wall to wall carpeting effectively eliminates any reverberation off the floor.
- Wall material. Brick and concrete block walls are efficient reflectors of sound, especially when they are painted. Walls made of gypsum, plaster, or wood tend to absorb sound more than brick and concrete, producing a more pleasing amount of reverberation.
- Windows and drapes. Glass is like a mirror to sound waves—the waves just bounce off the glass and back into the room. Large picture windows, mirrors, even large framed photos, can create excessive reverberation.
- Open areas leading to other rooms. Walls produce the most reverberation. The more a room is enclosed the more reverberation will result.

Using Non-parallel Walls

Reverberation is strongest with parallel walls—that is, one wall that faces another wall directly across the room. Look closely and you'll probably find the auditorium of your favorite movie theater uses angled walls. Most often, the walls are angled outward so that there is more distance between the walls in the back of the auditorium than there is in the front.

You may wish to consider replicating this design in your own home theater. You don't need to tear down walls; fabric or wood panels, positioned at an angle from the wall, will also work. Position the panels so that they act to decrease the width of the room nearest the screen.

Using Sound Absorbing Fabrics

Remember that a certain amount of reverberation is always welcome. It makes the sound seem more natural. On the other hand, if you find your home theater room exhibits too much reverberation, there are some simple steps you can take to reduce the unwanted sound reflections. For rooms that exhibit extreme reverberation, you may need to consider a number of the following.

- The least expensive reverberation fix is to add drapes over windows, or fabric over brick or concrete walls. You do not need to cover all the walls, nor do the walls need to be covered ceiling to floor. You can reduce the most objectionable reverberation by adding drapes at only the front and/or back of the room. Remember, you don't need windows behind the drapes! Drapes can also be hung from the ceiling to about waist height. Alternatively, an inexpensive cotton fabric (like that used for building stage scenery) can be applied either directly to the walls, or attached to a lightweight wooden framework, that in turn is secured to the walls. Cover the fabric with a suitable fabric paint if you wish, or leave it plain.
- The walls nearest the speakers can contribute a great deal to the overall reverberation in a room. By installing heavy fabric panels behind the speakers, you can reduce the echoes to a more reasonable level. Note that installing the fabric panels can reduce the overall sound level coming from the speakers, so take this into consideration. Also, placing the speaker too close to the wall and fabric panels can increase the low frequency response of the sound system, resulting in an overly "boomy" or "bassy" sound. Before making a permanent installation, you may wish to experiment with temporary fabric panels near the speakers to see how they affect the overall sound.
- A more costly approach to reducing reverberation is to add acoustic tile to the ceiling. This is especially helpful in rooms with a very high ceiling (12 to 14 feet). You can install a suspended ceiling that not only lowers the ceiling, but also uses acoustic tiles for greater sound absorption.
- People and furniture can absorb copious amounts of sound. Reverberation can be a problem if your home theater room is relatively empty of furniture and you're the only one watching the show. Consider adding plants, plush carpeting, or fabric panels on larger walls to compensate for an empty screening room.

Good and Bad Placement of Speakers

Your sound system is only as good as the placement of the speakers. Haphazard speaker placement can ruin an otherwise excellent sound system. Keep the following points in mind when placing speakers for best sound coverage. In all cases, at least two speakers are assumed to provide adequate coverage.

- Aim the speakers over the central portion of the audience. This usually involves turning the speakers in at the front corners of the room, as illustrated in *Figure 8-4a*.
- Avoid mounting the speakers so that they are parallel to any wall, as shown in *Figure 8-4b*. This can cause excessive reverberation.
- Avoid locating the surround speakers too far from the audience. Surround speakers can be placed to the side or rear of the audience, whichever is most convenient. If the speakers cannot be mounted on a wall, place them on sturdy pedestals beside or behind the seating area.

••• INSTALLING HOME THEATER •••

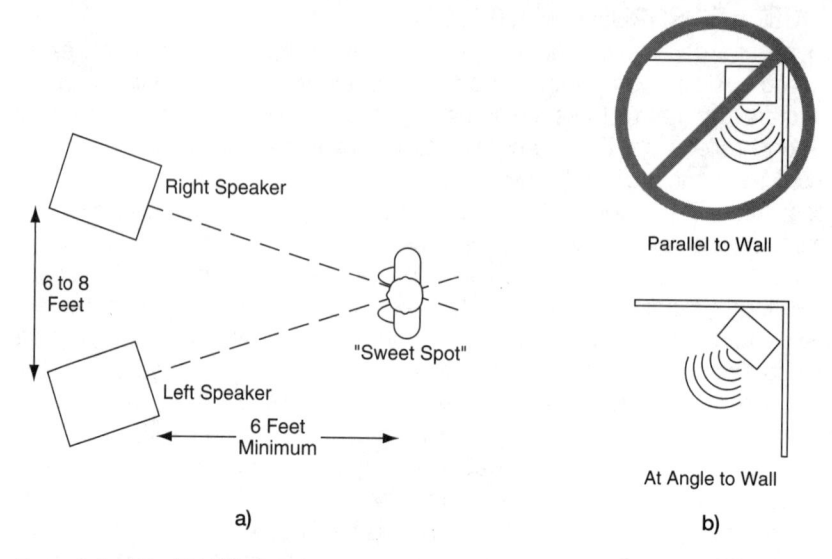

Figure 8-4. a) Position the right- and left-front speakers so they point into the audience, producing a "sweet spot" at the main seating area. b) Speakers placed parallel to a wall can create excessive reverberation. The speakers should be tilted at an angle from any nearby walls.

When first setting up your home theater system, it may be beneficial to temporarily place the speakers in the locations you feel are best suited for them. Try your sound system (preferably with people in the room) and sit at various places in the audience. Note any places where the sound is excessively loud or weak and reposition the speakers as necessary.

The traditional home theater speaker arrangement calls for two front channel speakers, a center front speaker, and two surround speakers (with an optional subwoofer). But the design of some rooms may preclude this ideal arrangement, and you should feel free to experiment with the placement of the speakers for best results. For example, if the room is very small, or if the front channel speakers must be placed closer than about 4-6 feet from one another, you may find a center speaker is not necessary. The right- and left-front channel speakers are close enough to provide a seamless sound stage. You can determine if the front center speaker is superfluous for your setup by disconnecting it or turning it off at the A/V receiver. Use your system for a few days without the front center speaker, then reconnect it. If you hear no difference, or the center channel speaker actually impairs the stereo imagery of the sound, leave it disconnected.

The walls are typical locations for the front and surround speakers. Rooms that lack walls are special problems in home theater design, because there's no place to mount the speakers. If the room for your home theater lacks walls, place the front channel speakers beside the TV screen, separating them by at least six feet. The surround speakers can be located on stands, placed either slightly behind or to either side of the audience. The surround speakers will not unduly distract the audience as long as they are at or above ear level, and their volume is not excessive.

8-8   Room Design and Ergonomics

••• INSTALLING HOME THEATER •••

The overall layout of your room will largely dictate how the speakers will be placed. A common approach is to place the left- and right-front speakers on shelves located to either side of the TV. This is acceptable as long as the speakers are not placed too high. You should avoid locating the speakers close to the ceiling, or excess bass is likely to result. In addition, the vertical height of all three front speakers (left, center, right) should be about the same (see *Figure 8-5*), and their distance to the audience should be consistent.

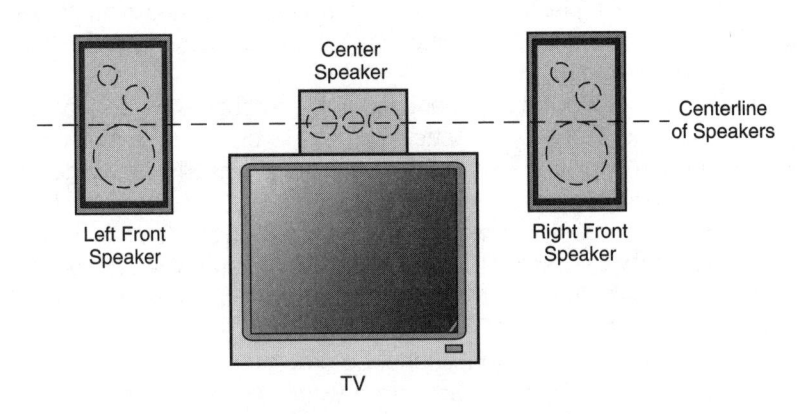

Figure 8-5. Our ears do not perceive minor differences in the vertical height of sound sources. Nevertheless, you'll achieve better overall sound quality if the all front speakers are on approximately the same horizontal plane.

Ideal Lighting for Home Theater

Good lighting is a must for any home theater. The light must not be too bright or directional, or the picture on the TV may not appear very bright (you may even be tempted to increase the brightness and contrast on the TV, which can damage the set over time). And the light must not be too dim, or you risk causing eye strain.

The ideal lighting for home theater is indirect or diffused lighting. Direct lighting is avoided because of the possibility of glare from the TV screen. You can test for glare by sitting at your normal spot in front of the TV and turning the set off. If you can distinctly see the glare or reflection from light fixtures behind or to your side, you need to reconsider the arrangement of the lighting in your room.

The easiest way to combat glare from lights is to turn them off. But that's not always a perfect solution. Others in the room may wish to have a light for reading, sewing, or knitting. And turning off all lights can lead to eye fatigue (you may have noticed that they seldom turn the lights all the way off in movie theaters). Lights that must be kept on when viewing your home theater should be relocated to eliminate glare.

With most modern TVs, glare is drastically reduced or eliminated if the light is more than about 20 degrees off axis to the screen. Of course, this applies to someone viewing the TV straight on. People sitting to either side of the screen may see objectionable glare from lights from the opposite side of the room. Experiment with the best location for light fixtures so that glare is reduced or eliminated for everyone, regardless of seating position.

Room Design and Ergonomics   8-9

••• INSTALLING HOME THEATER •••

If the full intensity of a lamp is not required, consider placing the fixture on a dimmer circuit. Dim the light to a warm glow while watching your home theater. You can purchase dimmer switches at most any home improvement store. You may also want to consider using a remote control dimmer, such as the Plug 'n Power® units available from Radio Shack. You can control the lighting in your home theater den, and even your entire house, without leaving the comfort of your easy chair.

If you have the benefit of installing new lighting for a dedicated home theater room, install recessed ceiling lamps (for use when watching TV) and indirect wall lighting (for general illumination). Place the ceiling lamps on a dimmer circuit so that you can reduce the intensity to a mild glow.

Glare can also occur from sunlight streaming through a window or door. Dimming the sun is not possible, so your better choice is to block the light as it comes into the house. Close the door or, if you want the outside air, install a heavy-duty metal screen door. The screen on the door will let in the air, but block much of the light. Put drapes on the windows to prevent sunlight from pouring through. You may wish to install two sets of drapes: thinner "shear" drapes that let in the sun when not viewing TV during the day, and thicker "blackout" drapes to block the light from entering the room and causing excessive glare on the TV screen.

In some instances the sunlight cannot be prevented from entering the room (example: light through a skylight, or light from a window in another room that does not or cannot have a drapery covering). In these cases, your best solution is to locate the TV so that the light does not reflect on the screen causing glare. For instance, if the sun comes in from the west, avoid placing the TV on an east wall, or it may get drenched in light. Positioning the TV on a south-facing wall is often a good compromise in these circumstances, since direct sunlight does not enter from a window with a northern exposure (for those living in the Northern Hemisphere, of course).

KEEPING YOUR HOME THEATER IN TIP-TOP SHAPE

This chapter concentrated on the science of acoustics and the ergonomics of good home theater room design. By now you probably have your home theater system installed and working. But to keep it working its best, you may want to consult the next chapter, "Home Theater Maintenance." It discusses important steps you can take to keep your system—from the A/V receiver to the satellite dish—in top working order.

Home Theater Maintenance

Odds are, your home theater system will work flawlessly for years. You can help ensure that it does so by performing some simple, routine maintenance to keep the system in top shape. With few exceptions (such as a VCR, antenna rotator, or satellite dish actuator), home theater systems contain no mechanical parts, so maintaining the components is usually a simple matter of keeping them clean. This chapter details the general cleaning procedures for home theater system components, including VCRs, video disc players, the cables and connectors, as well as video discs and tapes.

Read the Books!

An important measure in your efforts to minimize trouble with your home theater system is to *read the instruction manuals that came with your equipment!* This may sound obvious, but you'd be surprised how many home theater owners never take the time to thoroughly read the user's manuals. The equipment manufacturers include the manuals so that you can get the maximum enjoyment and longest life from your expensive purchases. Keep the manuals handy and refer to them whenever you have a question.

General Home Theater System Cleaning and Upkeep

The notion that "an ounce of prevention is worth a pound of cure" is certainly true for home theater components. Keeping the A/V receiver, TV set, VCR, and other components clean goes a long way toward preventing serious problems. The following sections describe general cleaning and upkeep of your system.

EXTERIOR CLEANING

At the top of your system maintenance list should be routine external cleaning of all components. This general discussion applies to the A/V receiver, TV set, VCR, video disc player, remote controls, and all other electronic components in your system.

Use a dry cloth to wipe off each component of your home theater system on a weekly basis. Make sure you get the back of the components as well, since this is where the cables connecting the system together are hidden from view. Don't use dusting sprays as these actually attract dust back onto your equipment. For hard to reach places, use a soft, sable painter's brush. Be sure to clean the ventilation slots as these are favorite hiding places for dust.

If you need to get rid of stubborn grime, apply a light spray of regular household cleaner (such as glass cleaner) onto a clean rag, then wipe with the rag. Never apply the spray directly onto a component, as the excess can run inside and might cause a short circuit. Be sure to check the label to make sure the glass cleaner is safe to use on plastics. Never use a petroleum- or acetone-based solvent cleaner. These can remove paint and melt the exterior plastic parts. Some plastics, when in contact with a solvent, let off highly toxic fumes that could seriously injure you.

CHECKING CABLES, CONNECTORS, AND JACKS

CONTINUITY AND SHORTS

Dirty or damaged cords and connectors can make your home theater system sneeze and wheeze. Inspect all power cords and plugs. Replace any cord that looks damaged. Inspect all cables and wires. Look for broken connectors. These, too, should be replaced or repaired when required.

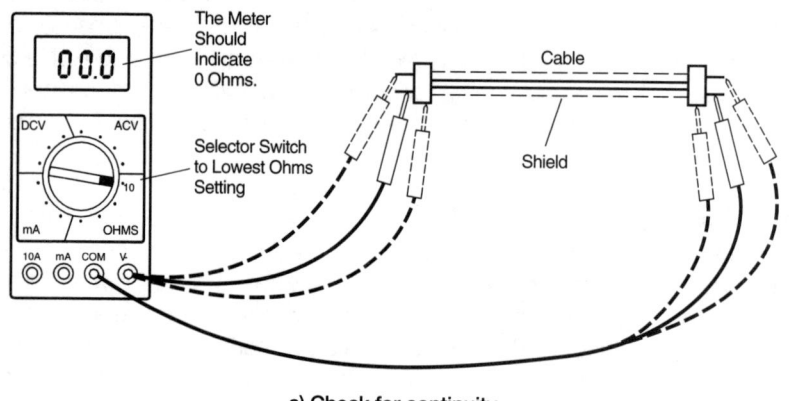

a) Check for continuity

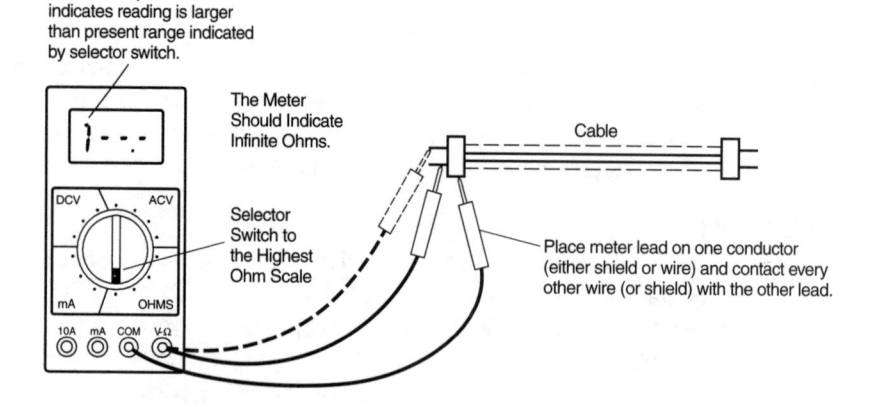

b) Check for shorts

Figure 9-1 Use a volt-ohm meter (VOM) to determine the continuity of a cable or wire. a) The cable is good if the meter reads approximately 0 ohms (continuity) when testing a conductor. b) The cable is shorted, and therefore unusable, if the meter reads 0 ohms across conductors.

9-2   Home Theater Maintenance

••• INSTALLING HOME THEATER •••

If you suspect that a cable may be bad—that its conductors are broken on the inside—try a replacement to see if the problem is solved. You also can use a volt-ohm (VOM) meter, as shown in *Figure 9-1a,* to test the continuity of the cable shield or its individual wires. Make contact with the leads of the meter to each cable conductor and take a reading. The meter should read 0 ohms indicating that there is a complete circuit through the wire shield.

Also, as shown in *Figure 9-1b,* check that the cable does not have a short between wires or between the shield and a wire by applying the test leads to each wire or shield and every one of its neighbors in the cable (be sure not to hold onto the metal probes of the test leads, you might get false readings). The measurement should read infinite ohms (open circuit).

OXIDATION

Video cables use un-tinned copper wire to carry the signal from place to place. Copper readily oxidizes (tarnishes) and the oxidation acts as a virtual shield to electrons. Whenever possible, use only sealed, molded cables. The all-in-one construction of the cable prevents oxygen from reaching the wire strands inside. Some high-end hookup cables are promoted as "oxygen free" to prevent signal-starving oxidation.

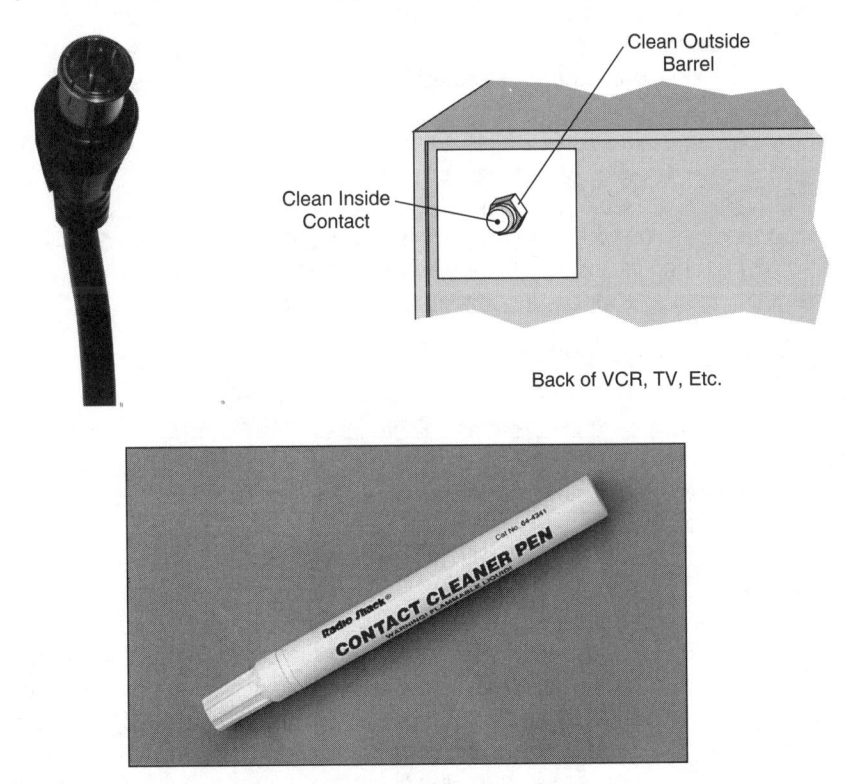

Clean Outside Barrel

Clean Inside Contact

Back of VCR, TV, Etc.

Figure 9-2. Repair damaged F-connectors by cutting off the old connector, trimming back the insulation and outer shield, and adding a new connector. Make sure the center wire is clean and free of oxidation. Use a suitable cleaning fluid to clean the cable connectors and plugs in your home video system. The cleaner in wipe-on "pen" form is the most convenient to use.

Home Theater Maintenance   9-3

If you make your own cables, you can help prevent oxidation by sealing the cable ends and connector. Apply a light coat of acetone-free silicone sealant to the joint between cable and connector. Don't get any of the sealant on the contacting points of the connector. Let the connectors dry and then plug them in place. Another method is to apply heat-shrink tubing over the joint. Before attaching the connector, slip a 3/4-inch length of heat-shrink tubing over the end of the cable. Secure the connector. Next, slide the tubing over the back end of the connector. Apply a small flame from a lighter or match to the tubing until it shrinks to fit. Avoid excessive heat or you'll melt the tubing. You can buy heat-shrink tubing at Radio Shack and most electronics outlet stores. It comes in various diameters, lengths, and colors.

F-connectors leave the center conductor of the coaxial cable bare. The conductor makes contact with a sleeve inside the mating connector of your VCR or TV. Because this center conductor is exposed and is made of copper, it is subject to oxidation. Check all F-connections in your video system and look for oxidized center conductors. Unoxidized copper looks bright and shiny like a new penny; oxidization darkens and dulls the metal.

The oxidation is nearly impossible to remove without harming the conductor; a better approach is to simply remake the connection. Cut off the existing F-connector, trim away the insulation and outer braid, and add a new F-connector, as shown in Chapter 5, *Figure 5-9*. Before adding the new connector, inspect the conductor to make sure the oxidation has not extended into the cable and affected more than just the exposed tip.

When you're satisfied that all cables and wires are in good shape, clean all the connectors to ensure good electrical contact, as shown in *Figure 9-2*. Suitable contact cleaner is available at Radio Shack, in aerosol spray or wipe-on "pen" form. You can also use cleaner designed for magnetic heads in video cassette recorders. It can be applied to the connector using a cleaning wand or cotton swab, and its low water content makes the cleaner evaporate completely in seconds.

Maintaining Your VCR

About 90 percent of all VCR malfunctions are mechanical, caused by little more than lack of basic cleaning and routine maintenance. You can do some of this rudimentary maintenance yourself and save some money. Of course, some VCR problems go beyond simple cleaning, but by taking care of the "little things" you can at least save the major problems for a qualified repair technician.

HEAD CLEANING

Like any electronic device that uses magnetic media, the record and playback heads in your VCR will eventually become dirty due to an accumulation of the oxide coating that sheds off of the tapes. The heads also may become dirty by direct contact with dirt, ash, grime, or other contaminate on the tape surface. Dirty video heads can create excessive drop outs—salt-and-pepper flecks that appear in the picture. Depending on the amount of drop outs, the flecks can be anything from a light shower to a heavy blizzard. There are two methods you can use to clean VCR heads: 1. use a tape cleaning cassette, or 2. clean the heads manually.

••• INSTALLING HOME THEATER •••

> *Important note:* After cleaning the VCR heads, wait at least five minutes before inserting a tape and playing it. This waiting period is important as it allows the cleaner to fully evaporate. If you play a tape while the heads are still wet, the cleaner may "melt" the oxide coating on the tape. The tape will be damaged, the heads will get dirty again, and this time the oxide residue will be even harder to remove. Throw the cleaning swab or chamois away after use. Although it may still seem clean, re-using it can impair the effectiveness of future cleanings.

CLEANING CASSETTES

The easiest, fastest, and safest way to clean video heads is with a commercially manufactured cleaning cassette, illustrated in *Figure 9-3*. Using it is simple: Apply cleaning fluid (some types only), pop the cassette into the machine, and "play" it for the recommended time (usually less than 20 seconds). Some head cleaning cassettes (particularly the "dry" kind that don't use a cleaning fluid) work by abrasive action, so you'll want to avoid using this type of cleaning tape until you really need it. It's possible to wear down a video head by using this type of cleaning cassette for an extended period of time, so follow directions carefully.

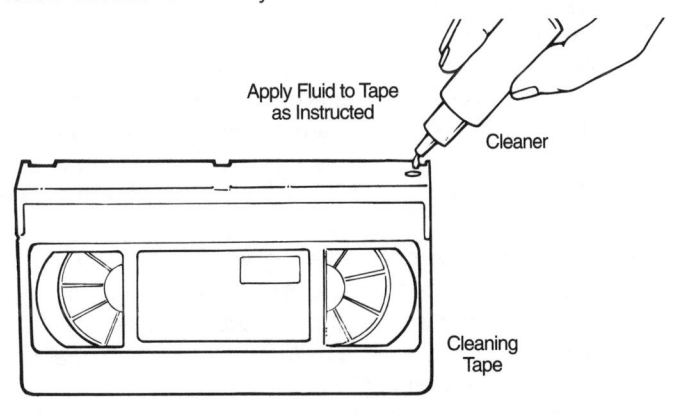

Figure 9-3. Use a cleaning cassette to clean the video and other magnetic heads inside your VCR. Some cleaning cassettes require the use of a cleaning fluid; others are used without a cleaning fluid.

MANUAL CLEANING

Manual cleaning of the video heads in your VCR requires you to remove the top cover of the machine, which may void its warranty. Proceed with caution and make sure you unplug the VCR from its power source before removing the cover.

Manual cleaning requires an alcohol-based cleaning solution and swabs that are available at Radio Shack and most electronics and video stores. *NEVER use cotton swabs as the lint can come off and become entangled in the head mechanism.* You may also use a piece of clean, virgin chamois, cut into a convenient 1-by-1 inch square. When cleaning, thoroughly wet the swab and GENTLY rub it sideways against the head. NEVER use an up-and-down motion—that can snap the heads right off. Most VCRs have several heads around the circumference of the drum; be sure you clean them all. Don't use a cleaning solution designed for audio heads. These sometimes have lubricants that aren't needed or desired for video heads.

FRONT PANEL CONTROLS

The front panel operating controls of the VCR seldom need preventive care, unless the machine has been subjected to heavy doses of dust, dirt, damp air, and cigarette smoke. You can clean the control panel on your VCR with a brush or a can of compressed air. Note that the control panel in most decks use sealed membrane switches, so the internal switch contacts are not affected by dirt (the actual switches are located behind the membrane switch buttons).

WHEN A VCR EATS A TAPE

If you've owned a VCR for any length of time, you may have lost a video tape when it became tangled inside the machine. Some VCRs are hungrier than others when it comes to "eating" tapes; some never do it, and others (particularly older models) consume more than their fair share. Your VCR may eat a tape for a variety of reasons, but the most common causes of eaten tapes are worn, dirty, or broken internal parts, particularly one of the belts used to move the tape during playback.

If your VCR is eating tapes and you have been able to remove the cassette, you should look for obvious signs of obstruction or a foreign object inside (use a flashlight) that could be tangling with the tape. Sometimes, small children stuff little objects in the convenient swinging door! If the obstruction or foreign object can be safely removed, do so. If you can't find anything wrong, or if you can't remove the obstruction or object, take your VCR in for repair. Using a video cleaning cassette will likely *not* fix the problem of a tape-eating VCR.

Should the tape become caught inside the VCR be very careful in extracting the loose ends, or you may cause irreparable damage. Do not pull on the tape, as this can cause further damage to both the VCR and the video tape itself. If necessary, you may need to remove the top cover of the VCR (unplug it first) to extract the tape. Taking your VCR apart may void its warranty, so check your owner's manual first. With the cover off, delicately remove the tape from the internal components of the VCR.

KEEPING VIDEO TAPES IN TOP CONDITION

Under normal use, your video tapes will last for years. Stored properly, they will produce about the same quality pictures years from now as they did when they were new. But all this assumes the tapes have been stored properly and handled with care. Proper storage and handling are keys to proper tape maintenance.

STORING VIDEO TAPES

Whenever possible tapes should not be stored flat. The strain caused by the reel and the weight of the tape itself can cause it to warp. For best results, the tape should be stored upright, so the weight of the reel and tape is distributed more evenly. When storing a videotape, place it spine up (another benefit: if the tape has a label, you'll be able to identify the tape easily).

There's no rule that says you must store videotapes in their cardboard or plastic dust jackets, though it is recommended that you do so. Some seasoned videophiles

••• INSTALLING HOME THEATER •••

throw out the tape covers of blank tapes because they get in the way. If you don't use covers, you should at least try to keep the tops of the tapes dust-free. If you live in a dusty area, and don't like constant housekeeping, consider buying a videotape storage cabinet. Make sure it lets you store the tapes the proper way (most do, but some poorly-designed cabinets overlook proper tape keeping).

VIDEOTAPES AND WEAR

One popular myth says you can use a tape only a certain number of times. This is true, of course, but that number is in the hundreds, and you'll likely never use a tape that much. Each time you play a tape, you scratch off a little bit of the oxide coating that stores the video and audio signals. At first, the amount isn't much, but after some time—say 60 to 70 playings—the results become visually obvious: the middle of the picture has flashing streaks of white. These streaks are known as dropouts and are caused by missing portions of magnetic oxide coating. Though the tape may exhibit some wear, you should not prematurely throw it away. Unless the tape is damaged or causes the video heads to become repeatedly dirty (and require cleaning), you should keep the tape and continue enjoying it.

You should be more wary of tapes that have been abused, or show obvious signs of damage. The most common causes of damage include:
- Mis-threaded tapes, where the tape jumps the threading hoops inside the deck and gets tangled in the mechanism.
- Smashed tapes, caused by the door of the cassette closing on tape that hasn't been rewound onto the reels, or that has become tangled inside the VCR.
- Stretched tapes, typically caused by a malfunctioning VCR.
- Scratches, usually caused by dirt or some foreign object in the VCR, or in the video cassette itself.

Damage doesn't always mean the tape is unwatchable. If the hurt is minor, the tape will still pass through the VCR without harming the heads. There will be a noticeable amount of picture noise for the duration of the damage, however. Severe damage, where the tape is badly mangled, torn, or stretched, means that it's unusable and should not be played.

MAINTAINING YOUR VIDEO DISC PLAYER

Mechanically, video disc players—either the digital DVD or analog laser disc varieties—are less complex than VCRs, so problems with a video disc player are less common. Nevertheless, it's a good idea to wipe off the top, sides and back of your video disc player to prevent dust build-up to help keep it in good working order.

OPTICS CLEANING

Video disc players use a sophisticated laser optics system to read the information stored on discs. The laser produces a thin beam of light that strikes against the disc, which then reflects the light to a photo sensor. As the reflected laser light changes in intensity a photo sensor inside the player converts the changes in the reflected light beam into an electronic signal. The signal from the sensor is then converted to audio and video. The information contained on the video disc is "stamped" onto a thin aluminum disc, which is then encased inside clear plastic to form the disc.

Home Theater Maintenance   9-7

Figure 9-4. Use a cleaning disc to clean the internal optics of your videodisc player. If required by the disc, apply suitable cleaning fluid prior to inserting the disc into the player.

The operation of the optics assembly in a video disc player is very exact. If the main optics of the player becomes dirty or greasy, playback is impaired. The picture and sound may become unstable or the entire disc may not be playable. The optics of the video disc player should be cleaned whenever there is playback trouble (clean the video disc first, as detailed later in this chapter, then clean the optics in the player if the problem persists). *You should not manually clean the lens of the player,* or the optics system could be damaged. The delicacy of the optics mechanism in a video disc player requires the use of a special cleaner. Cleaning discs *(Figure 9-4)* are available at Radio Shack and other stores. These cleaning discs use a pad or brush mounted on the disc that gently "wipes off" the lens. This pad or brush must be inspected periodically, and the cleaning disc replaced if the pad or brush is worn or dirty.

Using the cleaning disc is easy—simply insert the disc, pad, brush, or other cleaning surface face down. Press Play. Some cleaning discs require that you select a certain track on the disc (such as track 2 or 3). Let the disc play for 30 seconds, or for the amount of time indicated in the instructions that come with the equipment. When done, remove the cleaning disc from the player, and replace it into its holder. Note that the holder helps protect the pad or brush used on the disc. Do not store the cleaning disc without using its special holder.

TAKING CARE OF YOUR VIDEO DISCS

Video discs—either the 12-inch analog variety or the 4.7-inch DVD (digital) kind—are hearty and will last a long time if properly stored and handled.

STORING VIDEO DISCS

Like audio CDs, most digital video discs come in protective "jewel boxes" that provide convenient storage of the disc and liner notes, which provide information about the program stored on the disc. The jewel box helps keeps the disc clean and scratch-free. Always return a DVD disc to its jewel box when the disc is not in use. For the greatest protection, store the disc and jewel box in a CD drawer or cabinet.

••• INSTALLING HOME THEATER •••

The 12-inch analog video discs are more fragile, in part because they are larger with more surface area that can become dirty or scratched. And, analog video discs come in cardboard jackets like the old vinyl LP records, not sturdy plastic jewel boxes. A thin inner sleeve helps protect the disc against dirt and scratches. Always replace the disc into the inner sleeve when viewing is done, and slip the sleeve, open-side up, into the cardboard jacket (by inserting the sleeve open-side up into the jacket, there is less chance of the disc slipping out and dropping to the floor). Store the large 12-inch video discs *only* in an upright position, on edge. Never stack discs on top of one another— this can cause serious warping. Avoid placing discs on a shelf where they might lean against one another, putting pressure on other discs.

With either type, you must always handle the disc by the edges only. Oil from your skin can impair playback and attract dirt. Dirt causes scratches, and if the scratch is deep enough, the disc may be rendered unviewable where the scratch occurs.

CLEANING VIDEO DISCS

Most any commercial cleaner or cleaning kit for CDs can also be used on DVDs. The liquid-based cleaning kits are particularly easy to use: apply the cleaning fluid to the felt pad, insert the disc into the holder, then close the holder. Spin the disc a few times to distribute the cleaner. Remove the disc and allow it to dry, or pat it dry with a soft paper towel.

Discs also can be cleaned by hand. Use water or a very diluted mixture of ordinary household glass cleaner (important: make sure the glass cleaner is safe for use on plastics). Liberally apply the water or cleaner to the surface of the disc, and use a soft paper towel to dry it. Clean the disc from center to outside edge, *not in a circular motion.* This helps avoid scratches that can impair playback.

CLEANING THE REMOTE CONTROL

Most of your primary home theater components—including VCR, A/V receiver, television, VCR, satellite receiver, and video disc player—will come with a remote control. You may use some or all of the remote controls for your home theater gear, or you may choose to use a "unitized" all-in-one control that can operate all of the home theater components. Either way, you may find that the remote control(s) of your home theater system are subjected to considerable abuse, including being stepped on, spilled on, dropped, and more.

BASIC CLEANING

Cleaning the remote control is straightforward and easy. Start by wiping the exterior with a damp cloth to remove dirt and grime. Remote controllers are battery operated. Given the right circumstances, all batteries can leak, and the spilled acid can corrode the electrical contacts, impairing proper operation of the controller. If your remote control is operating sporadically, this may be the problem. If the batteries have leaked, or if the battery contacts have become even slightly dirty, clean them as follows.

1. Carefully remove the batteries and discard them. Immediately wash your hands to remove any battery residue.
2. Wipe off the battery residue inside the remote with a damp cloth, and remove as much of the excess from the contacts as you can.

3. Remove the remaining residue by rubbing the contacts with the tip of a pencil eraser (see *Figure 9-5*). Blow the eraser dust out of the battery compartment when you are through.
4. Insert a set of fresh batteries.

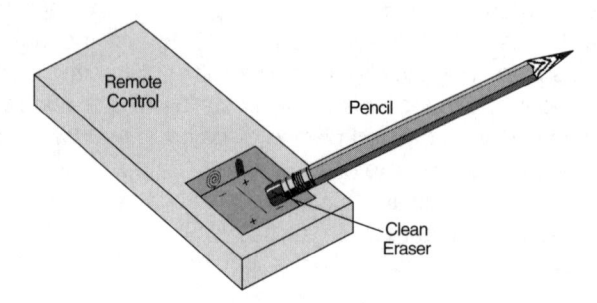

Figure 9-5. Use a new pencil eraser to clean the battery contacts inside the remote control. Be sure to blow out the eraser debris when done cleaning.

Note: It's a good idea to replace the batteries in the remote control every 6 to 12 months, even if they still seem to have "juice" remaining in them. Replacing the batteries on a regular basis will avoid unnecessary damage from leakage.

INTERIOR CLEANING

To clean the inside of the remote (to remove residue from spilled liquid, for example), you must disassemble it by removing the screws holding the two halves together. The following instructions aren't for everybody, so proceed with care. If you don't feel you have the skills or mechanical aptitude to disassemble the remote control, take the controller to a qualified repair technician. Be sure to remove the batteries from the remote control before disassembly.

Most remote controls use small screws, so you'll need a set of jeweler's screwdrivers to properly remove them. Some screws may be hidden in the battery compartment, under the manufacturer's label, or even beneath the thin metal cover on the front of the unit. With the screws removed, separate the two halves. Clean the inside of the controller with a soft brush or can of compressed air. If liquids have been spilled inside the controller, spray the inside of the controller with a suitable cleaner/degreaser, available at Radio Shack. (Note: Never use a lubricating cleaner, such as WD-40. This may cause permanent damage to your remote control.) The circuit board in most controllers is attached to the front panel and must be removed if you need to access the front switches. With the remote resting on a flat surface, remove the circuit board screws carefully and gently lift off the control board. Clean the switches and front side of the circuit board with a spray from the can of cleaner/degreaser.

Adjusting Your Television for the Best Picture

Most modern televisions are nearly automatic in self-adjusting for best picture quality. Some can use a reference signal provided by the broadcast station to ensure that the colors are rendered properly on the screen. Still, the quality of the picture on a TV screen is a subjective matter. What looks like a properly-adjusted picture to someone else may not be appealing to you. For this reason, all TVs provide a means to control (at the least) the brightness, contrast, color, and tint of the picture. These controls are especially important on older TVs that lack automatic color, tint, and other settings.

All TV sets come with a factory setting. However, you may want to alter this so that the picture is adjusted to your tastes. You'll probably also want to readjust the TV from time to time to ensure that it continues to deliver a good picture. Before you begin, read the owner's manual so that you clearly understand the adjustment procedure.

USING A BROADCAST PICTURE SOURCE

There are four main controls you'll want to "tweak." Some TVs come with additional controls that may need periodic adjustment as well. You may set these controls visually using a regular program (a news room broadcast is a good choice), or with a special setup tape or disc. Here's how to set up your TV using a broadcast picture source:

Brightness

The *brightness* control changes the relative lightness or darkness of the whole image. Some TV programs are meant to be lighter or darker than others (for mood), so you should set the brightness to achieve a "middle ground" that works for any channel or program you tune in. In general, the brightness control should be set so that the shadow and bright areas still have detail. Large areas of total black or white mean the brightness control is not set properly.

Contrast

The *contrast* control varies the ratio between lights and darks. When the contrast control is set too high, the picture is mostly light, with few shadows. Conversely, when the contrast control is set too low, the picture is mostly dark with few highlights. You'll want a good balance between light and dark. Adjust the contrast control—perhaps along with the brightness control—for a pleasing distribution of lights and darks.

Color

The *color* control varies the amount of color in the picture. You can turn the control all the way down for a black and white (or nearly so) picture. Or, you can turn the control all the way up for bold, overly-bright colors. Most people like to set this control just high enough so it doesn't distort any of the colors. If any of the colors smear—particularly the reds or oranges—turn the control down a bit.

Tint

The *tint* control varies the color tones in the picture. Most people adjust the tint control to achieve good "flesh tones" for people. This is a good starting point, but you may find it necessary to adjust tint and color to achieve a good balance.

ADDITIONAL CONTROLS

Sharpness

If your TV has a "sharpness" control, you may want to dial that to the mid-point and leave it there. The sharpness control attempts to make a picture softer or sharper, but usually ends up affecting its overall quality. However, if your TV is showing some age, increasing the sharpness control will compensate for loss of detail in the picture.

Convergence

Recall from chapter three that color picture tubes use three (sometimes two) electron guns to produce colors. These electron guns must be precisely aimed at the shadow mask and phosphors at the front of the tube; otherwise, the individual colors that make up the picture will not register properly. Instead of a solid white line, for example, you will see three separate lines: red, green, and blue. The "aiming" of the electron guns (convergence) is done electronically. Most modern, direct-view TV sets require little or no convergence adjustment, even after 10 or 20 years of use. But projection TVs require periodic convergence adjustments, especially if the set is moved.

Convergence is adjusted either with an internal and external adjustment, or only with an internal adjustment. *Internal adjustments should be made only by qualified technicians.* This adjustment requires that technician open the TV and adjust the picture while the set is on. *This is highly dangerous and poses a serious and lethal shock hazard. Therefore, do not attempt this adjustment yourself.*

Some projection units offer an external convergence adjustment that can be used for "fine tuning." You can safely adjust this control so that the three primary colors of the picture blend in as best as possible. A completely convergent picture is nearly impossible to achieve, especially on a very large screen projection TV. The center may be properly adjusted, but the colors may separate at the corners. Many users obtain the most acceptable results by setting the convergence for best picture in the top left, center, and top right, of the picture. Minor convergence error in the bottom of the screen is usually not as noticeable or annoying.

Solving Mysteries of Your Home Theater System

This chapter detailed the steps you can take to keep your home theater system in top working order. It described general upkeep and maintenance of the primary components of your system, including VCR and disc player. And it discussed ways to ensure that your television set is displaying the best pictures possible. In the next chapter, you'll read about what to do if you experience unusual problems with your home theater system, such as picture interference, buzzing or crackling in the sound, or a malfunctioning piece of equipment.

Troubleshooting Common Problems

Despite the best efforts to keep your home theater system in tip-top shape, problems and malfunctions can occur, causing poor sound or picture quality. But don't panic should your home theater system fail to work properly. In many cases, the real problem is minor and can be remedied in a matter of minutes.

This chapter presents troubleshooting tips for common problems with a home theater system. It is divided into several sections for your convenience. It begins with an overview of basic troubleshooting, then presents several component-specific troubleshooting tables. The chapter also provides additional information on troubleshooting and resolving problems caused by interference.

Basic Troubleshooting

Many problems with home theater systems are simple in nature and can be easily remedied. For example, bad reception may be fixed simply by fine-tuning your TV (some older sets have manual fine-tuning). Or, snow on the picture when playing a video tape can often be fixed by adjusting the tracking control on the VCR. Following is a quick summary of how to resolve basic home theater maladies:

PROBLEMS WITH THE TV AND RECEPTION:

- *Tune to the proper channel.* If you have a VCR, video disc player, or other source connected to your TV through the antenna-in connector, make sure the TV is dialed to the correct output channel for the video source. Usually, this is channel 3 or 4.
- *Adjust the fine tuning.* In this age of "automatic everything" it's sometimes easy to forget that not all television sets incorporate automatic fine-tuning. On sets that don't, you must adjust the fine-tuning control to bring the picture into sharpness. The same advice applies to an older VCR that likewise lacks automatic fine-tuning.
- *Watch the brightness and contrast controls.* If the picture looks overly dark or washed out, it may not be the fault of the TV or some other home theater component. Rather, it may be caused by a simple misadjustment of the set's brightness and contrast controls. On older TVs, these controls are often found inside a front panel. On newer TVs, the contrast and brightness controls are typically found in a special setup menu, accessible by pressing the Menu or Setup button on the TV or its remote control.
- *Vertical hold adjustment.* Most newer sets offer rock-steady pictures with automatic circuits to control the vertical hold of the picture. Older TVs lack these advanced circuits, and it may be necessary to manually adjust the vertical hold. This

is particularly true when playing older video tapes, or tapes recorded at the slower LP and SLP speed.
- *Fixing reception problems.* A bad picture (with or without bad sound) may be the result of interference. Interference can adversely affect a television picture quality when using cable, an outdoor antenna, a satellite dish, and even a VCR. See "Curing Television Interference," later in this chapter, for a more detailed description of TV interference, and how to deal with it.

PROBLEMS WITH THE VCR

- *Bad or broken cables.* Cables leading from or to your VCR should be in good working order. Broken or poorly made cables and cable connectors can impair signal quality. Repairing or replacing these faulty cables often remedies the most persistent video gremlins.
- *Bad tracking.* VCRs require a tracking adjustment so that the signal recorded on the magnetic tape can be played back accurately. On newer VCRs, this tracking adjustment is automatic. Occasionally, especially with older tapes or when using an older VCR, manual override adjustment of the tracking control is necessary.
- *Defective or damaged tapes.* Video tape has a finite life. After repeated viewings the tape itself may exhibit wear that can be seen (and heard) on the TV. Most often, worn tapes don't pose a hazard to your VCR, unless the tape is so old or deteriorated that it leaves bits and pieces of itself inside the VCR. The video and sound also may be impaired if the video tape has been damaged, which can occur if it is abused, played on a faulty VCR, or kept in a damp or hot place for an extended period of time.
- *Dirty video heads.* The magnetic heads in a VCR can become clogged with the magnetic coating from video tapes, and this can cause a poor picture quality and, in some cases, poor audio quality as well. The effect is seen as "snow" on the picture. To keep a top-quality picture, the VCR heads should be cleaned regularly, and must be cleaned when they become dirty.

PROBLEMS WITH THE VIDEO DISC PLAYER:

- *Scratches on the disc.* Video discs are made of plastic, which can be scratched. If the scratches are deep and large enough, they can impair playback. The effect is most noticeable on 12-inch analog video discs, and appears as jumps in the picture. On the smaller DVD digital discs, scratches can cause a momentary "freeze" of the video image, or even a temporary "blank-out" of the picture. Scratches can seldom be removed. The best course of action is to protect the disks to prevent scratching them in the first place.
- *Dirt and smudges on the disc.* A lesser evil is dirt and smudges on the disc that also can cause momentary playback problems. The dirt and smudges usually can be removed by cleaning the disc. Use a recommended CD/DVD cleaner with DVD discs. Use water (or diluted plastic/glass cleaner, if necessary) and a soft paper towel to clean 12-inch analog discs. Wipe the disc from the center to the outside edge—not in a circular motion.
- *Warped disc.* Video discs can become warped if exposed to heat or physical stress. A warped disc is almost always permanently ruined, and cannot be

played. Always be sure to store your discs in a cool place. Analog 12-inch video discs should only be stored in an upright position. Always store DVD discs in their jewel case.

PROBLEMS WITH THE SOUND:
- *No sound.* A/V receivers often are a complex mixture of buttons, dials, and indicator lights. Learning to use it properly may take time, and a thorough reading and re-reading of the owner's manual is strongly recommended. A common problem is no sound, which usually is caused by incorrect settings of the receiver's buttons. Make sure the program source (VCR, satellite receiver, etc.) is working, and recheck your switch selections
- *No sound through rear speakers.* The rear speakers contain the "surround" portion of the sound track. Not every program is encoded for surround sound, and little or no sound may come from the rear speakers. If you know the soundtrack contains surround information, be sure the volume level for the rear speakers has been properly adjusted, and that you have selected Dolby Surround or Dolby Pro-Logic mode on the A/V receiver.
- *Wrong sound.* A/V receivers are used to select the program source for viewing on the TV. Usually, picking a source (say, VCR-1) selects both the audio and video portion of the program. In some cases, the audio and video can be "split," as may occur if a combination video disc/CD player is connected to the A/V receiver. Depending on how you have the A/V receiver and disc player connected (as well as the design of the A/V receiver), you may need to first select Video Disc as the video source, then select CD as the sound source. Check the manual that came with the A/V receiver for details and wiring suggestions. You should also check the Tape Monitor button; pressing the button may result in no sound, or sound from an incorrect source.

DIAGNOSING POOR SOUND QUALITY

The bane of any home theater system is bad sound. Most people can live with a slightly fuzzy picture (for a limited time), but a raspy sound track grates on one's nerves. In this section you will learn about correcting the problems with sound quality caused by bad or loose connections, or interference from nearby electrical fields. These faults can cause annoying hum, static, hiss, and distortion.

HUM

A low frequency "hum" is one of the most common maladies of home theater sound systems. It is caused by the 60-Hz ac (alternating current) used to power the system. The volume of the hum can be soft or loud, depending on its source and where the noise enters your home theater system. By far, the most common cause of hum is poor grounding of the system components.

Hum also can be induced when the components in the system do not share the same electrical ground. This is common when the components are not plugged into the same outlet. Outlets separated by a distance may have different ground potentials. The ground at one outlet may actually be at 0.5 volts rather than 0.0 volts, and the ground at another may actually be at 3.7 volts. This difference can set up a *ground loop* condition, which can cause hum as well as heavy black bars on the TV screen. One way to ensure proper grounding between components is to plug them into the same ac receptacle or surge protector.

Another source of hum is caused by induction of ac signals from power cords into signal cables. If an ac power cord is positioned too close to a signal cable, the 60-Hz frequency in the power cord can be inductively passed into the signal cable. Hum caused by induction is typically caused by placing loops of ac cord and signal cable too close to one another. You can minimize or eliminate hum caused by induction by keeping the ac cords and signal cables physically separated, and by not coiling ac cords and signal cables.

BUZZ AND STATIC

Buzz is characterized as noise that has a steady frequency; that is, the buzzing is regular and doesn't change pitch. The frequency of the buzz depends on its source. It might be low- or high-pitched. Conversely, static is characterized by random noise. The static may range from an occasional "pop" once a minute to a constant crackle.

Buzzing

Buzz and static are often caused by the same thing — radio interference that is often created by an electrical appliance (often malfunctioning) operated in close proximity to the home theater components. For example, a steady buzz can be attributed to the radio frequency (RF) emissions from a ballast in a fluorescent light fixture. The unwanted RF signal (noise) travels through the air from the ballast to your home theater system.

Another source of a steady buzz is from dimmer switches. The switch uses an SCR (silicon controlled rectifier), which switches power to the load (such as an incandescent lamp) on and off many times each second. Depending on the age and condition of the dimmer switch, RF noise can result. You can readily isolate the problem to the lighting system by turning off the lights (if they are on dimmers, don't just dim them; turn the lights off completely). If the noise disappears, you know the problem is caused by a lighting fixture or dimmer switch.

Static

Whereas buzzing is a constant stream of noise, static is "aperiodic" (random crackling) and its sources are harder to track down. One typical source of static is noise caused by an automobile ignition system, especially an old car. The static is generated by the high voltage coil and the spark discharge within the engine. This "spike" of RF noise is transmitted to your equipment, where it can cause snaps, crackles, and pops in your sound system.

If it is not possible to repair the source of the static, or the source of the static cannot be found, you should consider better shielding in the wiring for your home theater system. Most static is picked up by the long lengths of signal wires, such as those between the VCR and A/V receiver. Higher-quality, shielded cable can often markedly reduce the effects of static and other RF interference.

DISTORTION

Sound is distorted when the shape of the output signal differs from that of the input signal. While the amplitude of the signals may be different due to amplification, the shape of the signals should be the same. The typical distorted output sounds "clipped"—the louder portions of the output are missing. The result of even modest distortion can be sound that is barely intelligible.

There are many types of distortion, and many causes for it. One cause might be an amplifier pushed beyond its design limits, trying to pump out more watts than it can really deliver. Reducing the output level of the amplifier restores the quality of the sound. Distortion also can be caused by an imbalance of input levels. Inputs are rated by their typical use and it is important to not "mix and match" inputs, or else distortion could result. Following are typical home theater input levels for audio signals:

Input type	Typical Level
Phono, magnetic	20 mV
Phono, ceramic	1.2 V
Line (VCR, CD, tape, etc.)	1.0 V

Notice the differences in output levels from a magnetic and ceramic phonograph needle cartridge. Phonographs that use a ceramic cartridge have a much higher output voltage. Most modern home stereo and A/V receivers lack a means to support ceramic cartridges, but a few have a switch that allows you to select Magnetic or Ceramic inputs. Note also that most A/V receivers have special circuitry for phono inputs and you should not connect any device to the phono inputs except a phonograph. The phono input incorporates "RIAA equalization" required to restore tonal balance when listening to records.

Curing Television Interference

As long as television signals are transmitted through the air or carried over a cable those signals will be susceptible to interference. TV interference can be caused by other radio signals, faulty wiring, nearby automobiles, even fluorescent lighting. While interference is a fact of television life, there's no need for you to be plagued by it.

Most cases of TV interference can be reduced or completely eliminated. Interference can be a persistent nightmare because it has many causes and symptoms. But by locating the source of your tormenting interference first, you can more easily and less expensively take steps to remedy it. Of course, no one solution will take care of all types of interference, and some forms of interference may be impossible to remove entirely. The sources of TV interference can be broken down into three main groups:
1. Interference from non-broadcast sources, like electric motors, car engines, and faulty fluorescent lights.
2. Interference from broadcast sources, like citizen band radios, AM radio stations, other television stations, and VCRs or cable boxes.
3. Interference from within the television set itself.

INTERFERENCE FROM NON-BROADCAST SOURCES

Your television is just one of several appliances you have in your house, and many of them are potential sources of interference. The most common is interference caused by ac and dc motors. Vacuum the rug or shave with an electric razor and odds are interference in the form of tiny sparkles and horizontal streaks appear on the screen. Usually, but not always, the visual interference is accompanied by a buzzing noise. Your best bet in stopping this type of interference is to install an ac line filter between the TV and the wall socket. The ac line filter also helps reduce or eliminate static caused by old or faulty light dimmer switches and fluorescent lamp fixtures.

Like the static on your sound system, another cause of non-broadcast interference is an automobile. Faulty ignition wiring can cause interference that you see as random sparkles or hear as an annoying whine that varies with the rpm of the car motor. The solution: repair the ignition wiring or install resistor-type spark plugs. If it's not your car, and you can't interest your neighbor in repairing it, you can try upgrading the antenna wire leading to your TV.

INTERFERENCE FROM BROADCAST SOURCES

Television interference caused by other broadcast sources, namely CB and amateur radio transmitters, AM and FM radio stations, and television stations, is harder to diagnose and treat than those caused by non-broadcast sources.

CB Radio

Interference from citizen's band (CB) radio takes two common forms: transmitter harmonic distortion and RF tuner overload. With transmitter harmonic distortion, your TV picture is filled with thin, slightly angled lines and garbled sound on channels 2, 5, 6, 9, and 10. The bad thing about this type of interference is that it can't be suppressed at your TV; you need the cooperation of the CBer. If the CB set is in your house, attach a low-pass filter, available at Radio Shack, to the transmitter. If the CB is somewhere in your neighborhood, you'll need to locate it and ask its owner to attach the low-pass filter.

RF tuner overload is caused by a strong CB signal invading your TV set. The effect is frightening: the picture may black out, the sound may be completely garbled, and the entire screen may be filled with wavy interference lines. This type of interference, which can occur on most any channel, can be corrected at your TV: Make sure your antenna system is in good working order, then add a CB high-pass filter or fixed attenuator to the antenna terminals. The attenuator cuts down the strength of the signals reaching your TV, so you need good reception to start with.

CB transmitters outfitted with illegal power amplifiers (anything over five watts) may totally overload your TV causing gross distortion on the screen. Often times, you can distinctly hear the CB operator through the speakers of your set. In some instances, the TV doesn't even need to be turned on; the signal radiated from the CB antenna is so strong that the wiring in your TV intercepts it. Such interference cannot be contained; the interference will only go away by stopping the transmissions.

Amateur Radio

Amateur radio sets are another potential source of TV interference. The symptoms are similar to CB transmitter harmonic distortion, except that the interference may appear on any channel (the exact channel depends on the frequency of the radio transmitter). A high-pass filter designed to block amateur radio interference can be added between your TV and antenna.

••• INSTALLING HOME THEATER •••

FM Stations

A somewhat common form of interference is caused by strong signals from a nearby FM radio transmitter. It's most often seen on channels 5 and 6, but can appear on any VHF channel. The visual effect is a series of wavy, grainy lines on the screen. The waves change with the beat of the music on the radio. Here are three possible solutions:
- Install a separate antenna for receiving FM radio broadcasts. Connect it directly to your stereo.
- Separate the UHF and VHF leads right at the antenna, instead of inside the house. Be sure to use a VHF/UHF signal splitter rated for outdoor use.
- Install an FM trap between your TV and the VHF antenna lead. If the trap has an adjustment knob, turn it until the interference is reduced or goes away.

Other TV Stations

Other television stations can cause the most disruptive form of interference. If you're located mid-way between two stations that broadcast on the same channel, both signals may overlap one another causing co-channel interference. In its mild form, co-channel interference looks like a series of dark horizontal lines — a Venetian blind effect. In its worst form, the interference displays a strobbing, sweeping pattern, much like windshield wipers.

Co-channel interference usually comes and goes with changes in the weather. The interference increases in warm weather when there's a change in the ionosphere; TV signals from even hundreds of miles away bounce off this high-altitude layer of the earth's atmosphere and right into your antenna. A better antenna, aimed precisely at the TV station you *do* want, is the best solution for co-channel interference.

Cross-modulation also is caused by two different TV channels interfering with one another. The interfering picture appears over the picture you want. Because the two pictures aren't synchronized, you often see a moving cross or bars on the screen, along with an irritating "herringbone" effect. As with co-channel interference, a better antenna can help solve the problem. If you're close to the broadcast towers, you may also need to install an attenuator to reduce the signal strength of the interfering channel.

INTERFERENCE FROM WITHIN THE TV SET

Some forms of interference are created inside your TV set. These are generally caused by a failed component or by improper service. Depending on the problem, you'll need to have the TV set repaired to eliminate the interference.

Common forms of TV-induced interference are color oscillator interference, channel 8 "tweet," and horizontal interference.
- Color oscillator interference appears as a series of steady, nearly vertical lines. It is caused by a poor connection in the color circuits of the TV. It also can be caused by an improperly installed television antenna lead-in.
- Channel 8 tweet (which occurs only on channel 8) appears as a series of swirling S's, and is caused by improper antenna connections and bad wiring or components inside your TV.
- Horizontal interference consists of one of more vertical lines, and is caused by failing components in the set. The position of the line (left, center, or right) in the picture generally indicates the faulty component.

CABLE TV INTERFERENCE

Just because you're on cable doesn't mean you won't be plagued by interference. In fact, depending on the quality of the equipment used in cable system, you may experience interference on some or all of the channels. This is especially true if the cable system is more than 10 or 15 years old and is not maintained on a regular basis.

Generally, interference generated within the cable TV system cannot be corrected at your TV. If you suspect that the cable system is at fault, contact the cable company by phone or letter. Describe the interference as best you can:

- What does it look like?
- What channels does it appear on?
- Does the interference also affect the sound?
- Does the interference change when you adjust the fine tuning, contrast, and color controls on your TV?
- Does the interference appear at all times during the day, or only at certain times.

Unless the cable company sends a repair technician to your home, you should not be charged for reporting the interference, even if the problem turns out to be caused by your equipment. Note that some cable systems are continually plagued by interference and you may have little recourse in fixing it.

Using Traps, Filters, and Other Devices

Many forms of externally-caused interference (that is, those that are created outside the TV set) can be minimized or eliminated with a trap, filter, or attenuator. These items are available at most electronics stores, including Radio Shack.

TRAPS REDUCE OR ELIMINATE UNDESIRABLE RADIO SIGNALS

Radio signals—whether from a CB or amateur radio rig, or an FM, AM, or TV station—are reduced or eliminated with the use of traps. Many traps, like an FM trap to reduce interference caused by strong FM radio signals, are tuned to just one specific frequency or a selected range of frequencies. Other traps are tunable: after installing it on your TV, you rotate a knob until the interference is reduced or eliminated. Traps are installed at the antenna terminals of your TV set, along with the antenna cable. For specific installation procedures, follow the instructions included with the trap.

FILTERS REMOVE UNWANTED NOISE

A filter is similar to a trap, except that it is generally used to mask random electrical noise, such as that caused by a faulty fluorescent light or the motor of your electric razor. Not all filters are the same and they're used in different ways. For example, a filter that reduces static caused by an automobile ignition is installed at the antenna terminals of the TV. A filter that reduces static generated by other electrical appliances in your house is installed at the ac plug of your TV. Be sure to purchase the proper filter for your needs.

ATTENUATORS REDUCE SIGNAL STRENGTH

Some signals can't be trapped or filtered. You use an attenuator to reduce the signal strength of all the TV channels if the incoming signal is overloading your TV (overload is characterized by a contrasty or negative image picture, plus excessive buzzing in the sound). An all-band attenuator reduces the strength of all channels equally.

If you're picking up two adjacent channels (like 5 and 6, for example) at the same time, you should use a tuned or tunable attenuator. A tuned attenuator reduces the strength of just one channel, such as channel 5, leaving the others relatively untouched. A tunable attenuator lets you dial the channel you want to reduce. The amount of attenuation can be either fixed or varied. A fixed attenuator reduces the signal strength by a set amount, generally 25 or 50 percent. A tunable attenuator reduces the signal from zero to up to 80 or 90 percent. Tunable attenuators can also be used to reduce ghosting—dial the knob on the attenuator to reduce the unwanted images.

Improving Your Antenna System

If you're not on cable, the first step in resolving interference is to improve your antenna system. On a clear day, and preferably with the help of a neighbor or relative (for safety purposes), carefully inspect the condition of your antenna. Has the antenna become corroded or rusted over the years? Look at the electrical connections between the antenna and lead-in cable. If they are corroded (and they usually will be), the connection should be cleaned and remade. Apply silicone sealant to protect the connection against the elements.

If you're experiencing lots of co-channel and cross modulation interference, a better antenna may be required. You need an antenna that is more directional—able to zero in on a channel and ignore others around it. A deep-fringe VHF/UHF antenna with yagi elements is a good choice. For more information on all types of TV (and other) antennas, read *Antennas* by A. Evans and K. Britain, published by Master Publishing Inc., available at Radio Shack stores (RS #62-1083).

Checklist for Removing Interference

When interference appears, don't rush out and buy $50 worth of filters, traps, and other accessories. Many times, the problem can be corrected without adding extra components to your video system. Follow this checklist when diagnosing and treating interference.

1. Dial to another channel. Some forms of interference appear only on certain channels.
2. Adjust the fine-tuning of the TV. You may be able to fine-tune the interference from the picture and sound. If your TV is equipped with an automatic fine-tuning control (AFC), turn it off first.
3. Inspect the wiring of your video system. Check the antenna, cable connections, and accessories, like signal splitters, A-B switches, and video routers. Poorly made accessories and bad connections can cause severe interference.
4. Bypass extra components in your system and connect the cable or antenna line directly to the TV. If the interference goes away, the problem is in your wiring, or within one of your accessories.
5. When possible, turn off all lights, radios, and non-necessary appliances in your home (this includes fans). If lights are on a dimmer circuit, be sure the dimmer is switched to OFF, not just LOW. If interference is intermittent, suspect appliances such as refrigerators and freezers.
6. Adjust contrast, color, and brightness controls on the TV to minimize the interfer-

ence. Some interference is not affected by adjusting the TV controls, so you won't see any difference on the screen.

TV INTERFERENCE CHART

Refer to this quick reference chart on pages 10-16 and 17 to help diagnose and fix specific types of television interference.

TROUBLESHOOTING CHARTS

The charts on the following pages will help you troubleshoot the most common problems with the major components of your home theater system. To use a chart, just locate the problem in the left column, and read the possible causes and solutions in the right columns.

ON WITH THE SHOW!

With your home theater is installed, calibrated, and working properly, it's time to sit back and enjoy your system. You'll probably want to show off (er, "demonstrate"!) your great new home theater system to family and friends. So buy or rent a favorite movie, pop it into the VCR or video disc player, and crank up the volume. Odds are, you and your guests will enjoy your home theater presentation better than the "real thing" at the local cineplex. And you can put all the butter you wish on your popcorn!

••• INSTALLING HOME THEATER •••

TELEVISION / MONITOR

Problem	Symptoms	Possible Causes	Possible Cures
TV will not turn on	Power On lamp doesn't light. TV will not operate	1. TV is unplugged 2. Switched outlet is turned off 3. Blown fuse or circuit breaker 4. Blown internal fuse	1. Plug into a known, good power outlet; check power plug that connects to the TV 2. Turn on switch, or plug into non-switched outlet 3. Reset circuit breaker or replace fuse 4. Replace fuse (may be internal)
TV turns on but no picture appears	Power On lamp lit but nothing appears on the screen	1. Brightness turned down 2. VCR or disc player in pause mode (or turned off) 3. A/V receiver switched to wrong input 4. TV faulty	1. Adjust brightness control 2. Put VCR or disc player into Play mode 3. Switch to proper input 4. Service TV
Poor picture quality	Picture looks grainy or snowy	1. Channel detuned 2. Faulty cabling 3. Poor reception 4. Signal too weak 5. Signal weakened by line loss	1. Fine tune until signal is clear 2. Inspect and repair 3. Adjust, repair, or replace antenna or cabling as needed 4. Use better antenna or use signal amplifier 5. Install amplifier if signal must pass through 100 feet or more of cable, or must be split
Ghosts in picture Note: Also see "Curing Television Interference" for more information	Images appear two or more times on screen	1. Antenna picking up reflected signal (ghost to right of main image) 2. Cable system cabling picking up signal from main TV station (ghost top left of main image)	1. Redirect antenna; use ghost eliminator 2. Use proper coax cable, keep lengths as short as possible; carefully fine-tune TV
Interference Note: Also see "Curing Television Interference" for more information	Picture (and/or sound) is distorted	1. Interference from electrical appliance motor 2. Interference from CB or other RF emitting device 3. Interference from other video device "mixing" with main signal 4. Signal too strong	1. Turn off appliance or install a power line filter 2. Install RF filter on antenna terminals of TV or VCR 3. Turn off interfering device; use coaxial cables (or repair cable if needed) 4. Use attenuator to reduce signal strength
Stretched or squeezed picture	Image appears distorted, as if stretched or squeezed	1. For stretched picture: Program source is displaying video in "widescreen" format intended for 16:9 television screen 2. For squeezed picture: Television is adjusted for 16:9 aspect ratio, but program material is not widescreen	1. Change program source to normal aspect ratio 2. Change TV to proper aspect ratio

Troubleshooting Common Problems   10-11

VCR

Problem	Symptoms	Possible Causes	Possible Cures
VCR will not turn on	Power On lamp doesn't light. VCR will not turn on or operate	1. VCR is unplugged 2. Switched outlet is turned off 3. Blown fuse or circuit breaker 4. Blown internal fuse	1. Plug into a known, good power outlet; check power plug that connects to the VCR 2. Turn on switch, or plug into non-switched outlet 3. Reset circuit breaker or replace fuse. 4. Replace fuse (may be internal)
VCR turns on but won't operate	Power On lamp lit but operating controls won't work	1. No tape in machine 2. VCR in timer mode 3. VCR paused 4. Condensation inside VCR 5. Tape jammed	1. Insert tape 2. Take out of Timer mode 3. Take out of Pause mode 4. Allow condensation to dry 5. Service VCR
Picture will not stabilize on playback	Rolling or jumping video; static lines on screen	1. Tracking off 2. Dirty or damaged tape 3. Dirty heads	1. Adjust tracking control 2. Replace tape 3. Clean heads
"Flagging" in picture	The top of the picture bends one way or another	1. Tracking off 2. Damaged (stretched) tape 3. Anti-copy signal on tape 4. VCR faulty	1. Adjust tracking control. 2. Replace tape 3. Replace or adjust TV for best picture 4. Service VCR
"Dropouts" in picture	Speckled dots flashing on screen	1. Tracking off 2. Old or damaged tape 3. Dirty video heads	1. Adjust tracking control 2. Replace tape 3. Clean heads
Noise in picture	Salt-and-pepper lines on screen; in extreme cases picture unviewable	1. Tracking off 2. Dirty video heads 3. Bad connections 4. VCR faulty	1. Adjust tracking control 2. Clean heads 3. Repair, clean, or replace 4. Service
VCR will not play tape	Picture unstable; tape stops after playing a short time	1. Tracking off 2. Defective tape 3. Condensation inside VCR 4. VCR faulty	1. Adjust tracking control 2. Replace tape 3. Allow condensation to dry 4. Service VCR
Poor sound	Hum, buzz, or pop in soundtrack Unstable soundtrack Distorted playback sound	1. Tracking off (especially when in hi-fi mode) 2. Cables defective 3. Improper use of audio inputs and outputs 4. Dirty audio heads	1. Adjust tracking control; switch from hi-fi to Normal audio 2. Repair or replace cables 3. Check and rearrange properly (refer to owner's manuals for hookup instruction) 4. Clean heads
No sound or wrong sound	VCR is working properly but there is no sound or the sound doesn't match the picture	1. Input selector on Line instead of VCR or Tuner 2. Audio cables disconnected 3. Audio and/or video cable plugged into Line input when receiving program off-the-air	1. Change input selector accordingly 2. Inspect and reconnect 3. Disconnect cable to receive on-air channel (required only on some VCRs)

VIDEO DISC PLAYER

Problem	Symptoms	Possible Causes	Possible Cures
Player will not turn on	Power On lamp is lit but operating controls won't work	1. No disc in machine 2. Player in timer mode 3. Player paused 4. Condensation inside player 5. Disk jammed	1. Insert disc 2. Take out of timer mode 3. Take out of pause mode 4. Wait for condensation to clear. 5. Service player
Player turns on but won't operate	Power On lamp lit but operating controls won't work	1. No disc in machine 2. Player paused 3. Disc defective 4. Controls jammed	1. Insert disc 2. Take out of pause mode 3. Replace with good disc 4. Service player
Picture jumps	The picture jumps and skips, either occasionally or constantly	1. Dirty disc 2. Scratched disc 3. Dirty player lens 4. Warped disc	1. Clean disc with approved cleaner 2. Inspect (cleaning may help) 3. Clean player lens with approved cleaner 4. Inspect disc and discard if necessary
Poor picture and sound	Picture and sound quality lower than usual	1. Dirty disc 2. Scratched disc 3. Dirty player lens 4. Disc recorded with Dolby Digital sound track	1. Clean disc with approved cleaner 2. Inspect (cleaning may help) 3. Clean player lens with approved cleaner 4. Switch to Dolby Digital mode (if A/V receiver so equipped) or switch to standard audio output on player and/or A/V receiver
Picture "freezes" (DVD discs only)	Picture appears to stop momentarily, then continue	1. Dirty disc 2. Scratched disc 3. Dirty player lens	1. Clean disc with approved cleaner 2. Inspect (cleaning may help) 3. Clean player lens with approved cleaner

REMOTE CONTROL(S)

Problem	Symptoms	Possible Causes	Possible Cures
Remote control doesn't work	Components do not respond when control buttons are pushed	1. Battery dead or weak 2. Strong light hitting remote sensor 3. Interference from another remote 4. Remote control defective	1. Replace battery 2. Shield remote sensor (on VCR, video disc player, cable box, etc.) from light source 3. Switch off other remote 4. Repair or replace
Remote control works occasionally	Controls function only occasionally; operation is sporadic	1. Battery weak 2. Strong light hitting remote sensor 3. Interference from another remote 4. Remote control is dirty	1. Replace battery 2. Shield remote sensor (on VCR, video disc player, cable box, etc.) from light source 3. Switch off other remote 4. Service or replace

A/V SYSTEM AND SPEAKERS

Problem	Symptoms	Possible Causes	Possible Cures
A/V receiver will not turn on	Power On lamp doesn't light; Receiver will not operate	1. A/V receiver is unplugged 2. Switched outlet is turned off 3. Blown fuse or circuit breaker 4. Blown internal fuse	1. Plug into a known, good power outlet; check power plug that connects to the receiver 2. Turn on switch, or plug into non-switched outlet 3. Reset circuit breaker or replace fuse 4. Replace fuse
A/V receiver turns on but no sound	Power On lamp is lit but operating controls won't work	1. Wrong input selection 2. No program source 3. Tape Monitor selected 4. Speaker switch in Off position 5. "B" speakers switched on when there are no "B" speakers connected (receiver inhibits output if no speakers are attached to the speaker terminals) 6. Speakers not connected	1. Select proper input (e.g. VCR-1 or Satellite) 2. Ensure program source is working and try again 3. Turn Tape Monitor switch off 4. Turn on speaker switch 5. Turn off "B" speaker switch 6. Inspect and reconnect
No or wrong audio	TV shows program video but wrong audio	1. Wrong input selection (e.g. CD mode when viewing DVD) 2. Loose or bad audio connections 3. Mismatched audio connections (e.g. VCR-1 audio plugged into Satellite jack)	1. Select proper input 2. Inspect and reconnect 3. Inspect and reconnect
Poor sound quality	Sound can be heard, but quality is below standard; sound is distorted	1. Bass and treble (or equalizer) controls not properly set 2. Loudness switch turned on 3. Volume output too high for speakers 4. Incorrect speaker connections (e.g. speakers connected to 4-ohm jacks rather than 8-ohm jacks) 5. Incorrect input connections (e.g. phonograph connected to VCR-1 input) 6. Video signal connected to audio jack	1. Inspect and adjust as needed. 2. Turn off Loudness switch 3. Reduce volume 4. Inspect and reconnect 5. Inspect and reconnect 6. Inspect and reconnect

••• INSTALLING HOME THEATER •••

A/V SYSTEM AND SPEAKERS (CONTINUED)

Problem	Symptoms	Possible Causes	Possible Cures
No sound from speaker	Sound can be heard from all but one speaker	1. Speaker wire bad or loose 2. Speaker wires shorted 3. Wrong speaker output mode (e.g. in 3-channel mode no sound is routed to rear speakers) 4. TV volume not turned up (when TV is used as the center speaker)	1. Inspect, repair, or replace as needed 2. Inspect and reconnect 3. Select proper speaker output mode 4. Adjust volume on television set
No stereo effect	A/V receiver is producing sound, but there is no stereo effect or it is diminished	1. Speaker wire bad or loose 2. Speaker wires shorted 3. Program is not in stereo 4. Wrong speaker output mode (e.g. 3-channel mode instead of Dolby Pro-Logic mode)	1. Inspect, repair, or replace as needed 2. Inspect and reconnect 3. Verify program contains stereo soundtrack 4. Select proper speaker output mode

TV INTERFERENCE CHART

Interference	Caused by	Appears as	Fixed by
Ac, dc, and universal motor noise	Electrical contact in brushes, commutators, or slip rings in electrical motors	Picture—Flashing specks or streaks Sound—Buzzing or crackling	Install ac line filter at television set
Fluorescent lamp noise	Faulty fluorescent lamp, ballast, or fixture	Picture—Flashing specks or streaks Sound—Buzzing or crackling	Replace lamp or ballast Repair faulty fluorescent lamp fixture Install ac line filter at television set
Dimmer switch noise	Old or faulty dimmer switch (generally when switch is set to low)	Picture—Flashing specks or streaks Sound—Buzzing or crackling	Replace dimmer switch Place ac line filter at television set
Automobile ignition	Faulty wiring in automobile; use of non-resistor spark plugs in automobile	Picture—Specks and streaks Sound—Buzzing or whining noise	Repair or upgrade automobile ignition system Replace antenna cable with shielded coaxial type Move antenna and cable away from traffic
Microwave oven; RF heating	Home or industrial RF heating units, such as microwave ovens	Picture—S-shaped moiré patterns in lower portion of screen; may roll vertically through picture Sound—None	Install high-pass filter on antenna leads of TV
CB transmitter harmonic distortion	CB radio	Picture—Diagonal lines on channels 2, 5, 6, 9 and 10; lines may move as CB operator speaks Sound—Garbled	Install a low-pass filter on CB radio
CB transmitter RF tuner overload	CB radio	Picture—Contrasty or completely dark picture Sound—Garbled	Re-orient antenna away from source of CB signals Use coaxial antenna cable Install high-pass CB filter at antenna terminals of TV set
Amateur radio fundamental and harmonic radiation	Amateur radio	Picture—Diagonal lines, fluctuates as radio operator speaks through microphone Sound—Garbled; may hear radio operator	Install high-pass filter at antenna terminals of TV set Install tuned trap at antenna terminals of TV set

Troubleshooting Common Problems

TV INTERFERENCE CHART (CONTINUED)

Interference	Caused by	Appears as	Fixed by
FM broadcast	Nearby FM stations	Picture—Grainy, fluctuating lines; fluctuating diagonal lines Sound—None or raspy noise	If the hi-fi is also connected to the TV antenna (through a suitable band splitter) install separate FM antenna Install FM signal trap at antenna terminals of TV
Co-channel interference	Two signals on same channel over-riding one another	Picture—Thick horizontal lines (Venetian blind); pulsating moving lines (windshield wiper) Sound—None or garbled	Orient antenna towards direction of desired channel Replace antenna with more directional model
Cross modulation	Two signals from adjacent channels over-riding one another	Picture—Herringbone effect; video cross point pattern slowly moving through picture Sound—None or slight raspiness	Orient antenna towards direction of desired channel Adjust fine-tuning Install tuned channel attenuator to block interfering channel (the interfering channel is the stronger one, and is the channel that appears over the one you want)
Color oscillator interference	Improperly connected antenna system; fault in color circuits inside the TV	Picture—Steady, slightly canted lines Sound—None	Repair or replace faulty antenna components Repair or replace faulty color circuits in TV
Channel 8 tweet	Bad antenna connections; fault inside TV	Picture—S-shaped herringbone pattern on channel 8 Sound—None	Repair or replace faulty antenna components Repair or replace faulty circuits in TV
Horizontal interference	Faulty TV	Picture—Thick vertical (sometimes bowed) black lines (usually appears only on weak channels) Sound—None	Service television set

••• INSTALLING HOME THEATER •••

CABLE, OFF-THE-AIR ANTENNA, OR SATELLITE RECEPTION

Problem	Symptoms	Possible Causes	Possible Cures
No reception	No picture (snow or "blue screen")	1. Cable box/satellite receiver not plugged in or turned on 2. Input cable disconnected 3. Cable faulty, e.g. conductors shorting, "open" conductor in cable, loose F-connector 4. TV dialed to wrong channel 5. No subscription for channel (cable or satellite systems)	1. Plug in and turn on 2. Inspect and reconnect 3. Replace or repair as needed 4. Dial TV to output channel of cable box or satellite receiver 5. Order channel from program provider
Snowy picture	Picture is visible, but reception is poor	1. Cable faulty, e.g. conductors shorting, "open" conductor in cable, loose F-connector, loose connectors 2. Antenna not pointed toward broadcast station or satellite 3. Signal strength reduced by long cable run or excessive splitting 4. Mismatched inputs/outputs	1. Replace, repair, or tighten as needed 2. Re-aim antenna 3. Add amplifier to boost signal 4. Use appropriate 75-300 ohm matching transformers, as needed
Scrambled channel (cable)	The picture is received, but appears scrambled (wavy lines, possibly no audio, etc.)	1. No authorization to receive the channel 2. Authorization lost (cable was disconnected, decoder box was turned off, etc.)	1. Order channel from cable company 2. Contact cable company to re-authorize channel
No authorization for channel (satellite dish)	The channel indicates "No subscription" (or channel is blank)	1. No authorization to receive the channel 2. Authorization lost (dish not pointed toward satellite, descrambler box is turned off, etc.)	1. Order channel from program provider 2. Contact program provider to re-authorize channel

10-18 Troubleshooting Common Problems

Glossary

A/B switch: A switch with two inputs and one output. Such a switch is used to select one of two signal sources for a component (TV, VCR, etc.) that can accept only one signal source at a time.
Absorption: The effect of a material that reflects only a portion of the sound waves striking it. Common absorptive materials include carpeting, acoustic tile, and drapes.
Acoustics: The science or study of sound.
AGC: An automatic gain control circuit built into a TV or VCR to automatically adjust the incoming signal strength to the proper level for display or recording.
Alternating current (ac): An electrical current that periodically changes in magnitude and direction.
Ambience: A surround or concert-hall sound.
Ampere (A): The unit of measurement for electrical current in coulombs per second. One ampere flows in a circuit that has one ohm resistance when one volt is applied to the circuit. See also *Ohm's Law.*
Amplifier: An electrical circuit designed to increase the current, voltage, or power of an applied signal. As used in sound systems, an amplifier (also a power amplifier) increases the current of an electrical signal from a microphone or other source and applies the amplified signal to one or more speakers so that it can be heard by a group of people.
Amplitude: The relative strength (usually voltage) of a signal. Amplitude can be expressed as either a negative or positive number, depending on the signals being compared.
Analog: Information presented in a continuous signal; as opposed to digital, which has two signal levels (on and off, or 1 and 0).
Antenna: As used in home theater systems, a wire or other conductive metallic structure used for receiving signals (TV, FM, etc.) broadcast through the air.
Aspect ratio: The ratio of height to width in a television or projection screen. Standard television has an aspect ratio of 4:3 (roughly 1.33:1); some "wide screen" TVs have an aspect ratio of 16:9 (about 1.78:1).
Attenuation: The reduction of an electrical signal, typically by some controlled amount.
Audio frequency: The acoustic spectrum of human hearing, generally regard to be between 20 Hz and 20,000 Hz.

Baffle: A piece of wood inside or outside a speaker enclosure to direct or block the movement of sound.
Balance: Equal signal strength provided to both left and right stereo channels.
Bandpass filter: An electric circuit designed to pass only middle frequencies. See also high-pass filter and low-pass filter.
Bandwidth: The amount or frequency range of a of signal that can pass through an electronic circuit in a given amount of time (usually one second). In general terms, the higher the bandwidth, the more information that can be conveyed; in picture and sound terms, this has the benefit of increasing audio and visual quality.
Balun: See *matching transformer.*
Baseband: Common term used to describe the separate audio and video signals used by VCRs, TV monitors, and many other video and audio components. Baseband audio and video signals are not accompanied by a radio modulation (RF) signal. See also *composite video.*
Bass: The low end of the audio frequency spectrum: approximately 20 Hz to 1000 Hz.

C-band: A commonly used band for commercial satellite television transmission and reception. The frequency range of the C-band is from 3.7 to 4.2 gigahertz.

Calibration: The adjustment of a television or audio system to ensure its output is within standard accepted levels. For a TV, calibration involves adjusting the color, hue, brightness, and contrast controls (and possibly others) to achieve the best possible picture. For an audio system, calibration involves adjusting the volume for each speaker and tonal equalization.

Cable: 1) One or more separately insulated wires bound together in or enclosed in a common protective covering. 2) A means of providing multiple television channels via closed circuit (i.e. not broadcast) from a common distribution point.

Cable-ready: A television or VCR equipped with a tuner capable of receiving channels normally used in a cable system.

Capacitance (C): The capability to store a charge in an electrostatic field. A capacitor is a device that stores electrical energy in the electrostatic field between two metallic plates. Most often used in home theater systems in speaker cross-over networks to pass/block certain audio frequencies. See also *inductance*.

Cassette: The two-reel plastic cartridge that contains audio or video magnetic tape.

Channel: 1) In audio, the separate signals of a home theater system. Includes front left, front right, front center, rear (in some systems separate left rear and right rear), and subwoofer. 2) In video, separate signals carried on an RF signal, used to isolate transmissions from one another.

Chrominance (C): The color component in a television signal. Chrominance represents the hue (color) and saturation (brightness level) of the picture. See also *luminance*.

Circuit: A complete path that allows electrical current from one terminal of a voltage source to the other terminal.

Coax: Short for "coaxial cable," a type of shielded cable used for video and some audio applications. It consists of a center conductor (which carries the signal) surrounded by insulation and a metal shield that guards the signal from interference.

Color temperature: A measurement of relative color difference, measured against an all-white reference. Color temperature is expressed in degrees Kelvin (K). A color temperature of 6,500 Kelvin is considered standard for color television sets.

Comb filter: An electronic device used in some TV sets that separates the *chrominance* and *luminance* components of a television signal.

Compact Disc (CD): A digital storage medium for sound, video, or data, typically measuring 4.7-inches. Information is stored on an aluminum (usually) disc that is encased in plastic. The information is decoded optically using a small laser light.

Component video: A video signal whereby the video signal is provided in three separate "channels": *luminance* (brightness), luminance and blue, and luminance and red.

Composite video: A combination of the various signals that are necessary for a TV or VCR to display or store a video image. The basic parts of a composite video signal are *luminance* (brightness), timing (synchronization), and—in the case of color signals—*chrominance* (color). Composite video is also called *baseband video*, or simply "video."

Convergence: In a TV set, the proper alignment of the red, green and blue beams from the electron guns in the television tube. Convergence error occurs when one or more beams are misaligned; the red, green, and/or blue colors are seen separately.

Converter: A device used to convert cable signals to a frequency and/or form suitable for reception on a standard television or VCR. The converter may include descrambling circuits, which restores the audio and video signals to their normal form.

CRT (cathode ray tube): A television picture tube.

Crossover network: An electric circuit or network that splits the audio frequencies into different bands for application to individual speakers.

Current (I): The flow of electrons measured in amperes.

Direct current (dc): Current that flows only in one direction.

Decibel (dB): A logarithmic scale used to denote a change in the relative strength of an electric signal or acoustic wave. The decibel is not an absolute measurement, but a ratio that indicates the relationship or relative strength between two signal levels

Decoder/descrambler: A device used with cable and satellite systems for restoring the audio and/or video signals to their proper form for viewing on a television set or recording on a VCR.

••• Installing Home Theater •••

Digital: A type of signal consisting of only levels, typically referred to as "off" and "on," or binary 0 and 1. Digital circuits are used in some home theater systems to process video and audio signals.

Digital Satellite System (DSS): A trade name for a consumer-oriented home satellite system. DSS is comprised of an 18-inch dish "antenna" for receiving signals from a satellite, and a receiver for tuning into the channels carried by a single satellite dedicated to relaying DSS programming. See also *HSS*.

Distortion: Any undesirable change in the characteristics of an audio signal.

Dolby®: Trademarked name for a variety of sound processing circuitry, including noise reduction (Dolby Noise Reduction) and surround sound (Dolby Pro-Logic, Dolby Digital). Dolby Surround is a system of encoding four channels of audio in a standard stereo soundtrack. Dolby Digital (also referred to as AC-3) is a system of encoding up to six discrete audio channels (five main channels and one subwoofer channel).

Driver: The electromagnetic components of a speaker, typically consisting of a magnet and voice coil.

Dynamic range: The range of sounds, expressed in decibels, between the softest and loudest portions that a system or speaker can reproduce without distortion.

DVD (Digital Video Disc): A compact disc video format capable of storing a minimum of 130 minutes of digitized picture and sound on a single disc.

Echo: A reflected sound wave of sufficient amplitude and delay to make it distinct from the original sound.

Equalizer: An adjustable audio filter inserted in a circuit to divide and adjust its frequency response.

Equalization: As used in audio, the adjustment of frequency response to tailor the sound to match personal preferences, room acoustics, and speaker enclosure design.

F-connector: The standard connector used with coaxial cable for video systems. The VHF and cable inputs and outputs on most TVs and VCRs also use F-connectors.

Farad: The basic unit of capacitance. A capacitor has a value of one farad when it can store one coulomb of charge with one volt across it. One farad is a considerable amount of storage; most capacitors are rated in millionths and even billionths of a farad.

Feedhorn: The part of a receiving satellite television antenna that collects the signals that are focused on it by the dish antenna's parabolic reflector element.

Fidelity: A measure of how true a circuit, amplifier, system, or subsystem reproduces its input signal.

Filter: An electrical circuit designed to prevent or reduce the passage of certain frequencies.

Frame: One complete video picture. In standard analog television, there are thirty video frames per second. Each frame consist of two half-frames, called *fields*.

Frequency: The number of complete cycles of a periodic waveform during one second; expressed in hertz (Hz).

Frequency response: The range of frequencies that are faithfully reproduced by a given sound system or speaker.

Full-range: A speaker designed to reproduce all or most of the sound spectrum.

Geostationary orbit: An orbit of space-borne satellites, approximately 23,500 miles above the earth. At that distance and position, the satellites revolve around the globe once in 24 hours, providing a stationary point of reference as seen from the earth.

Gigahertz (GHz): A measure of signal frequency equal to one billion cycles per second.

Ghost: A faint, double-image seen on a television screen caused by a reflection or echo of the transmitted signal.

Ground: Reference to a point of (usually) zero voltage, and can pertain to a power circuit or a signal circuit.

Harmonic distortion: Harmonic signals artificially added by an electrical circuit or speaker that are generally undesirable, expressed as a percentage of the original signal.

HDTV: High-definition Television, a standard for digital, high-resolution television transmission. Also sometimes referred to as DTV, for Digital Television.
Hertz (Hz): A unit of frequency equal to one cycle per second
High-pass filter: An electric circuit designed to pass only high frequencies. See also bandpass filter and low-pass filter.
Hiss: Audio noise that sounds like air escaping from a tire, typically caused by thermal noise generated by an electric circuit.
Home satellite system (HSS): A generic term for a complete system for receiving satellite television programming on the C- and Ku-bands. Also called "big dish," the satellite antenna is comprised of a parabolic dish measuring from 5 to 12 feet in diameter, feedhorn, and LNB. A receiver, located indoors, tunes into the individual channels available on one or more satellites positioned along the equator in space. See *DSS*.
Hum: Audio noise that has a steady, low-frequency pitch, typically caused by the effects of induction by nearby ac lines or leakage of ac line frequency into an amplifier's signal circuits.

Impedance (Z): The opposition of a circuit, speaker, wire, or cable to an alternating current. Note that impedance is the opposition to an alternating current (ac), where as resistance is the opposition to a direct current (dc). See *resistance*.
Inductance (L): The capability of an electronic component (typically a "coil") to store energy in a magnetic field surrounding it. It provides an inductive reactance which is an impedance to an ac current. Most often used in home theater systems in speaker cross-over networks to pass/block certain audio frequencies. See also *capacitance*.
Infrasonic: Sound frequencies below the 20 Hz threshold of normal human hearing. See *ultrasonic*.
Interference: Any external signal that causes disruption of the main signal. Interference can affect the audio or video portions of a television signal.
Interlaced video: A process whereby a video frame is composed of two separate fields. The first field contains the odd-numbered lines of video; the second field contains the even-numbered lines of video. Interlaced video is used to reduced flicker at lower frame rates (number of frames per second).
Intermodulation distortion (IM): The distortion that occurs when an electric circuit develops new frequencies from the ones being processed.

Kilohertz (kHz): A measurement of signal frequency equal to one thousand cycles per second.
Ku-band: A band for commercial satellite television transmission and reception, including Digital Satellite System (DSS). The total frequency range of the Ku-band is from 10.95 to 12.75 gigahertz. DSS uses a frequency range of 12.2 to 12.7 gigahertz.

LNB *(Low Noise Block-downconverter):* In home satellite TV systems, the component that receives the signals from the feedhorn, amplifies them, converts them to lower frequencies, and sends them to the satellite receiver.
LNBF *(Low Noise Block-downconverter Feedhorn):* Same as an LNB (see above), but with the addition of an integrated feedhorn,
LV: Short for LaserVision®, the trade name of analog optical laser video discs. Also referred to as laser discs, or LD.
Letterbox: A means of presenting a video program in "wide screen" format on a traditional 4:3 aspect ratio television screen. Instead of enlarging the center of the picture to full frame, the wide screen image is displayed horizontally centered on the screen. Black bars appear on the top and bottom where there is no picture.
Luminance (Y): The brightness (black and white) component of a video signal. See also *chrominance*.
Low-pass filter: An electric circuit designed to pass only low frequencies. See also bandpass filter and high-pass filter.

••• INSTALLING HOME THEATER •••

Matching transformer: A transformer that matches the output and input impedances. A typical application is to match the 300 ohm impedance of a television antenna to the 75-ohm impedance of coaxial cable. Also called a "balun," for balanced-unbalanced.
Megahertz (MHz): A signal frequency equal to one million cycles per second.
Midrange: A speaker designed to reproduce the middle frequencies of the sound spectrum, generally most efficient between about 1000 Hz to 4000 Hz.

NTSC *(National Television Standards Committee):* A group of businesses and engineers originally created to decide on early standards for color and black and white television in the United States. The NTSC system is also used in Japan, Canada, Mexico, and several other countries. Other television standards include PAM (used in most of Europe) and SECAM (used in France, parts of Africa, and Russia).
Noise: An unwanted sound or signal; interference.
Noise floor: The ambient sound and system noise below the program sound

Ohm (Ω): A unit of electrical resistance or impedance.
Ohm's law: A basic law of electric circuits. It states that the current (I) in amperes in a circuit is equal to the voltage (E) in volts, divided by the resistance (R) in Ohms. Thus, $I = E \div R$.

Pan and scan: A technique used when transferring a widescreen motion picture to the smaller *aspect ratio* of standard television. Only a portion of the picture is visible at any one time. If another portion of the picture must be shown, the frame is panned to that portion. See also *letterbox* and *widescreen*.
Peak: The maximum amplitude of a voltage or current.
PIP (Picture-in-Picture): A feature found in some TV sets and VCRs whereby one picture can be inserted over another, usually in a small frame in the corner of the main picture.
Polarity: The orientation of magnetic or electric fields. The polarity of the incoming audio signal determines the direction of movement of a speaker cone.
Power: The time rate of doing work or the rate at which energy is used. One watt of electrical power is the use of one joule of energy per second. Watts of electrical energy equals volts times amperes.

Resolution: The sharpness of a television image. Resolution is measured in different ways. Horizontal resolution is measured along a horizontal plane, and determines the overall sharpness of the picture. Different TV and video systems offer different horizontal resolutions (also called bandwidth). Vertical resolution is measured along a vertical plane. The vertical resolution is specified by television standards, and does not differ between video systems.
Resonance: In audio systems, the tendency of a speaker to vibrate most at a particular frequency; sometimes referred to as natural frequency.
Resistance (R): In electric or electronic circuits, a characteristic of a material that opposes the flow of direct current. Resistance results in loss of energy in a circuit dissipated as heat. See also *impedance;* note that impedance is the opposition to an alternating current (ac), where as resistance is the opposition to a direct current (dc).
RF (radio frequency) signal: The high-frequency, modulated signal transmitted through the air or through a cable that carries the information signals, such as baseband audio and video. Also called the carrier frequency.
RMS: An acronym for root mean square. The RMS value of an alternating current produces the same heating effect in a circuit as the same value of a direct current.
Reverberation: Echo-like repetition of an original sound signal. Reverberation can be caused by physical boundaries, such as sound bouncing off a wall, or created electronically to add "depth" to a sound signal.

S-video: A video signal whereby the *chrominance* (color) and *luminance* (brightness) components are provided in separate cables.

Satellite dish: A parabolic shaped antenna used to receive satellite signals. The dish itself is merely a reflector used to collect and concentrate the weak signals received from the satellite; the "active" portion of the antenna is the *feedhorn* and LNB.
Satellite receiver: Also called an IRD (for Integrated Receiver/Decoder), a tuner capable of receiving and converting satellite signals. Many satellite receivers also incorporate signal descramblers to restore video and audio signals to their proper form.
Signal: The desired portion of electrical information.
Signal-to-noise (S/N): The ratio, expressed in dB, between the signal (sound you want) and noise (sound you don't want).
Sine wave: The waveform of a pure alternating current or voltage. It deviates about a zero point to a positive value and a negative value. Audio and video signals are sine waves or combinations of sine waves.
Sound pressure level (SPL): The loudness of an acoustic wave stated in dB that is proportional to the logarithm of its intensity.
Sound spectrum: The range of sound frequencies discernable by the human ear, generally accepted as 20 Hz to 20,000 Hz. In actuality, sound waves can exist with frequencies just above zero Hz (infrasonic or subsonic), and well above the range of human hearing (ultrasonic).
Splitter: An electric device used to produce two or more outputs for one RF input. Splitters are commonly available with two or four outputs, thereby splitting the RF signal two or four ways.
Static: Random noise in a sound system due to atmospheric or manmade electrical disturbances, such as lightning.
Subsonic: See infrasonic.
Subwoofer: A speaker specifically designed to reproduce extremely-low frequency signals, from about 50 Hz to near zero Hz. Because of the very-low frequencies involved, subwoofers often create a "sensation" of sound and air movement, rather than an audible tone.
Surround: 1. As used in home/public theater sound systems, a generic term for the speakers placed to the rear of the listening audience, to provide sound ambience. 2. As used in speakers, the outer suspension of a speaker cone.

Three-way: A type of speaker systems composed of three ranges of speakers housed in the same enclosure, specifically a tweeter, midrange, and woofer. See also two-way.
THX®: A set of surround sound standards developed by Lucasfilm designed to ensure high-quality reproduction of the sound track.
Total harmonic distortion (THD): The percentage, in relation to a pure input signal, of harmonically derived frequencies introduced in the sound reproducing circuitry (including the speakers).
Treble: The upper end of the audio spectrum, usually reproduced by a tweeter.
Transient response: The instantaneous change in an electronic circuit's output response when input circuit conditions suddenly change from one steady-state condition to another.
Tweeter: A speaker designed to reproduce the high or treble range of the sound spectrum, generally most efficient from about 4000 Hz to 20,000 Hz.
Two-way: A type of speaker systems composed of two ranges of speakers housed in the same enclosure, usually a tweeter and midrange, or midrange and woofer. See also three-way.

UHF: Ultra High Frequency. TV channels 14 through 69.
VHF: Very High Frequency: TV channels 2 through 13.
VHS: The most popular home VCR format, pioneered by JVC.
Ultrasonic: Sound frequencies above the 20,000 Hz threshold of normal human hearing. See also infrasonic.

Watt: A unit of electrical power.
Woofer: A speaker designed to reproduce the low frequencies of the sound spectrum, generally most efficient from about 20 Hz to 1000 Hz.

INDEX

Acoustics: 8-3
Antennas: 2-1
   Accessories: 6-10
   Indoor: 6-9
   Off-the-air: 6-6
      Dipole: 6-6
      Yagi: 6-7
   Pointing: 6-12
   Satellite: 6-1~2
      DBS: 6-2
      DBS installation: 6-2
      HSS: 6-2
      HSS installation: 6-4
   Troubleshooting: 10-9
Aspect ratios: 3-8
A/V Receiver: 1-1,2; 4-3, 5-9
   Channel inputs: 4-4
   Channel wattage: 4-3
   Sound modes: 4-3

Bandwidth: 2-8
Bell, Alexander Graham: 4-1

Cable TV: 2-1~2
   Outlet plate: 7-10
Cables: 5-1, 5-4
   Keep separate: 5-13
Chrominance: 2-11; 3-7
Coaxial: 5-2, 5-17
Comb filter: 3-7
Components:
   Placement of: 5-3
   Hookups: 5-4
Component Video: 2-12; 3-7
Connectors:
   Broken: 7-3
   Continuity: 9-2
   F-type: 5-14
   Shorts in: 9-2
   Tools for: 5-15
   TV types: 3-4
   Types of: 2-12~14

DBS Satellite: 2-5
Dickson, William: iv
DigiCipher: 6-5
Digital TV: 2-8; 3-7

Dolby:
   Digital: 4-5
   Pro Logic: 1-4; 4-4
   Surround: 4-4
DVD (digital video disc):
   Disc cleaning: 9-9
   Disc storage: 9-9
   DIVX format: 2-7
   Optics cleaning: 9-8
   Players: 2-6~7; 5-7
   Regional Codes: 2-7
   Troubleshooting chart: 10-13
DTS surround sound: 4-7

Edison, Thomas A.: iv

HDTV: 3-7~8
Home Satellite System (HSS): 2-5

Feedhorn: 6-4
Frequency: 8-3

Impedance: 5-16
Installation tips: 7-11
Interference: 10-1
   Troubleshooting chart: 10-17

LePrince, Louise: iv
Lighting: 8-9~10
LNB: 6-4
LNBF: 6-3
Luminance: 2-11; 3-7

Maintenance: Chapter 9

Oxidation: 9-3

Remote control cleaning: 9-9
Resistance: 5-16
Reverberation: 8-4
RGB video: 2-12

Safety: 1-10
Satellite TV: 2-4; 5-6
Sonic hole: 4-2

Index   I-1

Sound:
    Absorption: 8-5, 7
    Decibel (dB): 8-3
    Frequency: 8-3
    Reverberation: 8-4, 5
    Troubleshooting: 10-9
    Volume: 8-2
Sound level meter: 1-9; 5-18
Speakers:
    Calibrating output: 5-18
    Center channel: 4-2; 4-8; 5-13
    Connecting: 5-9
    Front: 4-7
    Mounting: 7-8
    Placement: 8-7
    Surround (rear): 4-8
    Subwoofer: 4-9; 5-11
Stereo modes: 4-3
Stereo sound: 4-1
Surround sound: 4-1
S-Video: 2-11; 3-7

Television sets:
    Adjusting: 9-11
    Aspect ratios: 3-8
    Digital TV: 3-8
    Direct view:3-1; 3-10
    Features: 3-5
    HDTV: 3-8
    Large screen: 3-10
    Lenses: 3-13
    Monitor: 3-2
    Projection: 3-2; 3-10
    Scan rate: 2-8
    Size: 3-3
THX surround: 4-7
Troubleshooting:
    Charts: 10-11
    DVD problems: 10-2
    Sound: 10-35
    TV reception: 10-5
    VCR problems: 10-2

VCR: 2-2~4
    Cleaning: 9-4
Video:
    Bandwidth: 2-8
    Resolution: 2-8
    Scan rate: 2-8
    Signals: 2-10
VideoCipher: 6-5

Video signals:
    Component: 2-10
    Composite: 2-10
    Direct: 2-9
    Propagation: 6-8
    RBG: 2-11
    RF: 2-9
    Strength: 6-8
    S-video: 2-10
Videotape: 9-6
Volume: 8-2
VOM meter: 1-9; 9-2

Wire: 5-1; 7-3
Wiring:
    Attic: 7-5
    Baseboard: 7-7
    Basement: 7-4
    Outside: 7-8
    Planning: Chapter 7
    Raceways: 7-9
    Tools needed: 7-1
    Under carpet: 7-7